MISCELLANEOUS PUBLICATIONS
MUSEUM OF ZOOLOGY, UNIVERSITY OF MICHIGAN

The publications of the Museum of Zoology, University of Michigan, consist of two series—the Occasional Papers and the Miscellaneous Publications. Both series were founded by Dr. Bryant Walker, Mr. Bradshaw H. Swales, and Dr. W. W. Newcomb.

The Occasional Papers, publication of which was begun in 1913, serve as a medium for original studies based principally upon the collections in the Museum. They are issued separately. When a sufficient number of pages has been printed to make a volume, a title page, table of contents, and an index are supplied to libraries and individuals on the mailing list for the series.

The Miscellaneous Publications, which include papers on field and museum techniques, monographic studies, and other contributions not within the scope of the Occasional Papers, are published separately. It is not intended that they be grouped into volumes. Each number has a title page and, when necessary, a table of contents.

A complete list of publications on Birds, Fishes, Insects, Mammals, Mollusks, and Reptiles and Amphibians is available. Address inquiries to the Director, Museum of Zoology, Ann Arbor, Michigan.

List of Miscellaneous Publications on Reptiles and Amphibians

No. 8. The amphibians and reptiles of the Sierra Nevada de Santa Marta, Colombia. By Alexander G. Ruthven. (1922) 69 pp., 12 pls., 2 figs., 1 map ... $1.00

No. 29. A contribution to a knowledge of the herpetology of a portion of the Savanna Region of Central Petén, Guatemala. By L. C. Stuart. (1935) 56 pp., 4 pls., 1 fig., 1 map $0.50

No. 47. A contribution to the herpetology of the Isthmus of Tehuantepec. IV. An annotated list of the amphibians and reptiles collected on the Pacific slope during the summer of 1936. By Norman Hartweg and James A. Oliver. (1940) 31 pp. $0.35

No. 49. Studies of Neotropical Colubrinae. VIII. A revision of the genus *Dryadophis* Stuart, 1939. By L. C. Stuart. (1941) 106 pp., 4 pls., 13 figs., 4 maps $1.15

No. 50. A contribution to the knowledge of variation in *Opheodrys vernalis* (Harlan), with the description of a new subspecies. by Arnold B. Grobman. (1941) 38 pp., 2 figs., 1 map $0.35

No. 56. Taxonomic and geographic comments on Guatemalan salamanders of the genus *Oedipus*. By L. C. Stuart. (1943) 33 pp., 2 pls., 1 map $0.35

No. 61. Home range, homing behavior, and migration in turtles. By Fred R. Cagle. (1944) 34 pp., 2 pls., 4 figs., 1 map $0.35

No. 69. The amphibians and reptiles of Alta Verapaz, Guatemala. By L. C. Stuart. (1948) 109 pp., 10 figs., 1 map $1.50

No. 76. Studies of the black swamp snake, *Seminatrix pygaea* (Cope), with descriptions of two new subspecies. By Herndon G. Dowling. (1950) 38 pp., 6 figs., 1 map $1.25

No. 91. A brief review of the Guatemalan lizards of the genus *Anolis*. By L. C. Stuart. (1955) 31 pp. $0.50

No. 94. The anatomy of the head of *Ctenosaura pectinata* (Iguanidae). By Thomas M. Oelrich. (1956) 122 pp., 59 figs. $1.85

No. 96. The frogs of the hylid genus *Phrynohyas* Fitzinger, 1843. By William E. Duellman. (1956) 47 pp., 6 pls., 10 figs., 4 maps $0.70

No. 97. Variation and relative growth in the plastral scutes of the turtle *Kinosternon integrum* Leconte. By James E. Mosimann. (1956) 43 pp., 1 pl., 24 figs. $0.75

No. 101. A biogeography of reptiles and amphibians in the Gómez Farías Region, Tamaulipas, México. By Paul S. Martin. (1958) 102 pp., 7 pls., 7 figs., 4 maps $1.50

No. 110. Descriptions of tadpoles of Middle American frogs. By Priscilla Starrett. (1960) 38 pp., 1 pl., 33 figs. $1.10

No. 111. A systematic study of the lizards of the *deppei* group (Genus *Cnemidophorus*) in México and Guatemala. By William E. Duellman and John Wellman. (1960) 80 pp., 1 pl., 16 figs. $1.75

MISCELLANEOUS PUBLICATIONS
MUSEUM OF ZOOLOGY, UNIVERSITY OF MICHIGAN, NO. 111

A Systematic Study of The Lizards of The *Deppei* Group (Genus *Cnemidophorus*) in Mexico and Guatemala

BY

WILLIAM E. DUELLMAN
AND
JOHN WELLMAN
Museum of Natural History, University of Kansas

ANN ARBOR
MUSEUM OF ZOOLOGY, UNIVERSITY OF MICHIGAN
FEBRUARY 10, 1960

CONTENTS

PAGE

INTRODUCTION 5
  Acknowledgments 6
  Materials and Methods 6

RESUMÉ OF CHARACTERS STUDIED 7
  Size 7
  Scutellation 10
  Coloration 12
  Sexual Dimorphism 15

SYSTEMATICS OF THE deppei GROUP 16
  Definition 16
  Historical Review 16

KEYS TO THE SPECIES AND SUBSPECIES OF THE deppei GROUP 16
  Key to the Identification of Juveniles 17
  Key to the Identification of Adults 18

ACCOUNTS OF THE SPECIES AND SUBSPECIES 19
  Cnemidophorus deppei Wiegmann 19
  Cnemidophorus deppei deppei Wiegmann 24
  Cnemidophorus deppei infernalis, new subspecies 32
  Cnemidophorus deppei cozumelus Gadow 35
  Cnemidophorus lineatissimus Cope 38
  Cnemidophorus lineatissimus lineatissimus Cope 41
  Cnemidophorus lineatissimus exoristus, new subspecies 44
  Cnemidophorus lineatissimus duodecemlineatus Lewis, new combination 48
  Cnemidophorus lineatissimus lividus, new subspecies 50
  Cnemidophorus guttatus Wiegmann 54
  Cnemidophorus guttatus guttatus Wiegmann 57
  Cnemidophorus guttatus immutabilis Cope 61
  Cnemidophorus guttatus flavilineatus, new subspecies 65

DISCUSSION 69

SUMMARY 72

LITERATURE CITED 73

ILLUSTRATIONS

PLATE
(Plate I opposite page 80)

Holotypes of *Cnemidophorus deppei infernalis; Cnemidophorus lineatissimus exoristus; Cnemidophorus lineatissimus lividus;* and *Cnemidophorus guttatus flavilineatus.*

FIGURES IN TEXT

FIGURE   PAGE

1. Terminology of dorsal pattern in *Cnemidophorus* of the *deppei* group 13

2. Map showing distribution of races of *Cnemidophorus deppei* 21

3. Geographic variation in the number of dorsal granules in *Cnemidophorus deppei* 22

4. Geographic variation in the number of femoral pores in *Cnemidophorus deppei* .. 23

5. Map showing distribution of the races of *Cnemidophorus lineatissimus* 38

6. Geographic variation in the number of dorsal granules in *Cnemidophorus lineatissimus* 40

7. Geographic variation in the number of femoral pores in *Cnemidophorus lineatissimus* 40

8. Map showing distribution of the races of *Cnemidophorus guttatus* 55

9. Geographic variation in the number of dorsal granules in *Cnemidophorus guttatus* 56

10. Geographic variation in the number of femoral pores in *Cnemidophorus guttatus* 57

11. Ontogenetic change in color pattern in *Cnemidophorus deppei deppei* and *Cnemidophorus deppei infernalis* 75

12. Ontogenetic change in color pattern in *Cnemidophorus deppei deppei* (Lerdo de Tejada, Veracruz) and *Cnemidophorus deppei cozumelus* 76

13. Ontogenetic change in color pattern in *Cnemidophorus lineatissimus lineatissimus* and *Cnemidophorus lineatissimus exoristus* 77

14. Ontogenetic change in color pattern in *Cnemidophorus lineatissimus lividus* and *Cnemidophorus lineatissimus duodecemlineatus* 78

15. Ontogenetic change in color pattern in *Cnemidophorus guttatus guttatus* and *Cnemidophorus guttatus immutabilis* 79

16. Ontogenetic change in color pattern in *Cnemidophorus guttatus flavilineatus* .. 80

A SYSTEMATIC STUDY OF THE LIZARDS OF THE
DEPPEI GROUP (GENUS *CNEMIDOPHORUS*)
IN MEXICO AND GUATEMALA *

INTRODUCTION

LIZARDS of the genus *Cnemidophorus* long have been a thorn in the sides of herpetologists. Cope (1900: 569) stated: "The discrimination of the North American species of this genus is the most difficult problem in our herpetology." In a similar vein Gadow (1906: 287), who made the first serious attempt to unravel the systematics of the Mexican species, stated: "Most of the 'species' are so plastic, so variable, that they may well drive the systematist to despair. Not two authorities will, nor can, possibly agree upon the number of admissible species." After such comments from the "past masters," is it any wonder that most modern systematists have refrained from working with *Cnemidophorus*?

Probably the most confusing thing about these lizards is the ontogenetic change in color pattern, which through its variety of manifestations runs the gamut in convergence and parallelism. Gadow (*supra cit.*) presented a detailed account of the then accumulated knowledge of the Mexican species and attempted to show the ontogenetic change in color pattern in various populations. His classification of the group was based on a non-geographic concept of subspecies and hence is useless. Richard Zweifel recently (1959) made the first real contribution to the systematics of this genus within the last half century. By using size of the dorsal granules (a character first employed by Lowe and Zweifel, 1952), he was able to show that the supposedly highly variable *Cnemidophorus sacki* in western México actually included three species.

The senior author became aware of a systematic problem in the *deppei* group of the genus when, in 1956, he found two distinct forms of *"deppei"* in the Tepalcatepec Valley in Michoacán. During that field season much material from various parts of the range of the group was added to that which he had collected in Michoacán in 1951 and 1955. In 1958 we were able to visit Michoacán again. Among other things, we concentrated on obtaining series of *Cnemidophorus* and ecological data on the various forms. Upon comparing this material with series from diverse parts of México, it became obvious that the *deppei* group was badly in need of revision. Fortunately, with the exception of *cozumelus*, we have had field experience with all of the species and subspecies and have seen live juveniles, subadults, and adults of both sexes of all of the forms. Because

* Contribution No. 45 from the Department of Biology, Wayne State University, Detroit 2, Michigan,

we had not seen living specimens from Central America nor had a knowledge of the lizards in the field, we decided to limit the study to populations in México; however, on the insistence of L. C. Stuart, Guatemalan specimens have been included.

Therefore, the purpose of this paper is to review the species of the *deppei* group of *Cnemidophorus* and to present a classification of them, using some taxonomic characters which previously have not been applied.

ACKNOWLEDGMENTS

For permission to examine specimens or for information concerning specimens in their care we are grateful to Doris M. Cochran, United States National Museum (USNM);[1] Alice G. C. Grandison, British Museum (Natural History) (BMNH); Norman Hartweg and Charles F. Walker, University of Michigan Museum of Zoology (UMMZ); Robert F. Inger and Hymen Marx, Chicago Natural History Museum (CNHM); Hobart M. Smith, University of Illinois Museum of Natural History (UIMNH); Heinz Wermuth, Zoologisches Museum, Berlin (ZMB); and Richard G. Zweifel, American Museum of Natural History (AMNH).

Norman Hartweg, L. C. Stuart, and Richard G. Zweifel have provided us with innumerable suggestions and much valuable information, which has helped, we hope, to make this study more nearly complete and our interpretations more sound.

Many of the specimens used in this study were collected by field companions of the senior author; Ann S. Duellman, Richard E. Etheridge, Fred G. Thompson, and Jerome B. Tulecke deserve special thanks for their efforts.

None of the field work would have been possible without the cooperation of the Museum of Zoology; for their continuous support the senior author wishes to thank Norman Hartweg and T. H. Hubbell. Field work also has been supported in part by grants from the American Philosophical Society, the National Academy of Sciences, and the Graduate School of Wayne State University.

MATERIALS AND METHODS

During the course of this study we have examined 2302 specimens, of which 1236 have been studied in detail. We have not examined all available specimens, but we have attempted to study all specimens from critical areas. The type specimens of *Cnemidophorus deppei, duodecemlineatus,* and

---

[1] The museum numbers of all specimens referred to in the text and contained in the list of locality records are preceded by abbreviations of the museums; the University of Kansas Museum of Natural History is abbreviated UKMNH.

guttatus were not examined, but series of topotypes of each form were studied.

Throughout the text means are given in parentheses after the observed ranges; if the standard error of the mean has been calculated, this is given after the mean, i.e., 91–110 (102.5 ± 0.81). The specimens examined are listed chronologically after museum abbreviations according to alphabetical arrangement of localities in their political units. Methods of counting scales and stripes are discussed in the following section. The synonymy given in the account of each subspecies includes all of the names and combinations of names applicable to that form; no attempt has been made to make the synonymies complete, but important references since Burt's (1931) revision are included.

RESUMÉ OF THE CHARACTERS STUDIED

The classification of *Cnemidophorus* has been in such a chaotic state that we feel a discussion of the taxonomic characters used in this study will be of benefit to other workers. For additional comments on these characters and others in *Cnemidophorus sacki* and related species, see Zweifel (1959).

SIZE

The species, and to a lesser extent the subspecies, of *Cnemidophorus* in this group differ in the maximum size attained by the adults. So far as known, males always reach a greater size than females. (The largest specimen of the small series of *C. deppei cozumelus* is a female.) The maximum known snout-vent lengths for each form are given in Table I. Size is a difficult character to use. Many series, particularly those collected in the dry

TABLE I

MAXIMUM KNOWN SNOUT-VENT LENGHTS IN THE *deppei* GROUP

(Measurements in mm.)

Form	Male	Female
deppei deppei	93	87
deppei infernalis	84	75
deppei cozumelus	77	83
lineatissimus lineatissimus	96	79
lineatissimus exoristus	98	82
lineatissimus duodecemlineatus	92	72
lineatissimus lividus	106	89
guttatus guttatus	129	105
guttatus immutabilis	145	115
guttatus flavilineatus	113	93

TABLE II
VARIATION IN THE NUMBER OF DORSAL GRANULES IN THE *deppei* GROUP

Population	N	Range	Mean	SD	SE
deppei deppei					
Entire sample	422	90–142	110.6	11.70	0.57
Guerrero Coast	86	90–114	99.4	4.86	0.54
Oaxaca: Tehuantepec	52	91–110	102.5	5.81	0.81
Chiapas: Tonolá	25	110–123	117.4	5.53	1.12
Chiapas: Pijijiapan	14	110–141	119.5	7.76	1.33
Chiapas: Soconusco	63	115–142	129.5	6.93	0.88
Chiapas: Barra de Cahuacán	19	108–136	115.7	6.02	1.25
Guatemala: Pacific Coast	24	103–122	113.1	4.56	0.93
Chiapas: Grijalva Valley	22	106–123	113.6	4.46	0.95
Guatemala: Cuilco Valley	27	96–112	105.7	4.31	0.83
Guatemala: Motagua Valley	30	100–121	107.8	4.26	0.78
Northern Veracruz	14	91–112	102.6	9.60	2.57
Southern Veracruz	31	91–121	101.6	7.08	1.27
Veracruz: Lerdo de Tejada	19	109–124	114.7	4.74	1.08
Campeche: Ciudad Carmen	10	111–119	114.7	2.45	0.77
deppei infernalis					
Entire sample	251	88–120	99.3	8.82	0.56
Tepalcatepec Valley	233	91–120	101.4	8.87	0.58
Upper Balsas Basin	18	88–110	96.1	6.34	1.73
deppei cozumelus	30	103–118	110.6	6.75	1.23
lineatissimus lineatissimus	55	110–133	120.0	5.44	0.73
lineatissimus exoristus					
Entire sample	57	108–140	122.4	7.86	1.04
Tepalcatepec Valley	45	108–135	121.9	7.37	1.10
Pacific slopes, Michoacán	12	109–140	124.5	9.52	2.75
lineatissimus duodecemlineatus					
Entire sample	69	125–142	132.9	3.85	0.47
Nayarit	32	126–142	134.1	3.73	0.66
Jalisco	36	125–141	131.8	3.68	0.61
lineatissimus lividus					
Entire sample	53	126–164	148.0	9.01	1.24
Boca de Apiza to La Placita	26	137–164	147.6	7.57	1.51
Maruata and Pómaro	19	126–164	148.4	8.98	2.06
Río Nexpa to Playa Azul	8	129–155	143.3	7.41	2.97
guttatus guttatus	25	184–208	199.3	5.53	1.11
guttatus immutabilis					
Entire sample	98	153–198	176.8	15.30	1.54
Guerrero	68	155–194	175.8	10.06	1.22
Oaxaca	30	153–198	179.4	12.52	2.29
guttatus flavilineatus					
Entire sample	40	142–184	158.8	9.66	1.53
Pacific Coast	19	150–184	165.4	7.48	1.72
Cintalapa Valley	19	142–158	151.3	4.26	0.98

TABLE III

VARIATION IN THE NUMBER OF FEMORAL PORES IN THE *deppei* GROUP

Population	N	Range	Mean	SD	SE
deppei deppei					
Entire sample	417	29–44	36.5	2.66	0.13
Guerrero Coast	85	32–42	35.9	2.09	0.23
Oaxaca: Tehuantepec	51	32–44	37.1	2.49	0.35
Chiapas: Tonolá	25	35–42	38.6	2.46	0.49
Chiapas: Pijijiapan	14	33–42	36.8	2.16	0.57
Chiapas: Soconusco	62	32–42	37.9	2.58	0.33
Chiapas: Barra de Cahuacán	18	33–41	36.6	2.05	0.47
Guatemala: Pacific Coast	24	33–40	36.4	1.87	0.38
Chiapas: Grijalva Valley	21	33–41	37.3	2.54	0.55
Guatemala: Cuilco Valley	27	33–40	36.2	2.19	0.42
Guatemala: Motagua Valley	30	33–44	37.3	2.55	0.47
Northern Veracruz	15	31–39	35.4	4.28	1.10
Southern Veracruz	31	29–35	33.7	2.09	0.38
Veracruz: Lerdo de Tejada	19	31–42	34.0	2.37	0.54
Campeche: Ciudad Carmen	10	32–35	33.1	1.00	0.31
deppei infernalis					
Entire sample	252	31–44	35.6	2.29	0.14
Tepalcatepec Valley	234	31–43	35.6	2.14	0.14
Upper Balsas Basin	18	31–44	35.4	3.48	0.82
deppei cozumelus	29	32–38	34.7	1.46	0.27
lineatissimus lineatissimus	54	29–39	33.3	2.38	0.32
lineatissimus exoristus					
Entire sample	57	32–47	38.8	3.65	0.48
Tepalcatepec Valley	45	32–46	38.4	3.21	0.48
Pacific slopes, Michoacán	12	34–47	39.0	2.16	0.63
lineatissimus duodecemlineatus					
Entire sample	68	28–38	33.3	2.07	0.25
Nayarit	32	28–37	32.4	1.96	0.35
Jalisco	36	30–38	34.2	1.81	0.30
lineatissimus lividus					
Entire sample	53	32–48	37.9	2.79	0.38
Boca de Apiza to La Placita	26	33–43	36.1	2.91	0.57
Maruata and Pómaro	19	34–46	38.8	2.70	0.62
Río Nexpa to Playa Azul	8	38–48	41.9	2.08	0.77
guttatus guttatus	25	38–48	44.2	3.74	0.75
guttatus immutabilis					
Entire sample	97	34–52	42.8	3.65	0.37
Guerrero	68	34–48	41.9	3.31	0.40
Oaxaca	29	38–52	45.2	3.25	0.60
guttatus flavilineatus					
Entire sample	40	32–45	38.2	2.65	0.42
Pacific Coast	19	33–45	39.0	2.65	0.61
Cintalapa Valley	19	32–42	37.2	2.39	0.55

season, contain only juveniles and subadults. However, with the apparent
exception of *cozumelus,* we have seen several adult males of each form.
Because so many specimens have either broken or regenerated tails, the
length of the tail and its relative length to the body is of limited
use. Although we have not used the length of the tail in diagnosis, in the
description of each form we have presented measurements of tail length and
tail/body ratios for a small number of large males.

SCUTELLATION

Of the many features of scutellation only four appear to have any
taxonomic significance within this group; these are discussed in the individual species accounts, and with certain other scale characters in the
analysis below.

DORSAL GRANULES.—Lowe and Zweifel (1952) and Zweifel (1959) have
stressed the importance of the size of the dorsal granules in distinguishing
populations of *Cnemidophorus.* The present study includes the use of this
character in the *deppei* group for the first time.

The flanks and dorsum of *Cnemidophorus* are covered with small granular scales. The size of these granules may be expressed by counting the
number of granules around the middle of the body (not including the
enlarged ventral plates). These granules are arranged in rows, which
sometimes are irregular. Although eye strain is inevitable, with the use of
a binocular dissecting microscope accurate counts of the dorsal granules
can be made. For this study counts were made on about 1200 specimens.
Several individuals were counted repeatedly to ascertain the amount of
error; this was found to average less than 3 per cent. Specimens that have
been coiled or poorly preserved so that the skin is folded on the flanks are
especially difficult to count. Collectors should keep in mind that specimens
of *Cnemidophorus* that are preserved with the body straight and filled with
preservative will be much easier to identify.

In the *deppei* group there is great variation in the size (therefore, the
number) of the dorsal granules (Table II). Generally the large species have
smaller granules than the small species; thus, the large *guttatus* has as many
as 208 dorsal granules at midbody, whereas the much smaller northern race
of *deppei* has no more than 120 and may have as few as 88. Throughout
most of the range of the group sympatric species show no overlap in the
number of dorsal granules. For example, in southern Veracruz *guttatus*
has 184 to 208 granules at midbody, and *deppei* has 91 to 121. Although
two forms often have an overlap in the number of dorsal granules, only
in the Tepalcatepec Valley in Michoacán does this occur in sympatric
populations of two species. However, here there is a significant difference

in the average number of granules; *deppei* has 91 to 120 (101.4 ± 0.58), and *lineatissimus* has 108 to 140 (122.4 ± 1.04).

Unlike *Cnemidophorus sacki* in western México, which has a total range of only 91 to 120 dorsal granules and a range of means of the samples from 101.5 to 109.5 (Zweifel, 1959: 107), the species in the *deppei* group show considerably more geographic variation in this character (Table II). Therefore, certain populations have been united under one specific name with some degree of hesitancy; for example, of the four races of *lineatissimus*, those from Colima and the Tepalcatepec Valley resemble one another more closely than either does the population on the coast of Michoacán and that in Jalisco, and Nayarit. The geographic variation in dorsal granules and other characters of scutellation are discussed in detail in the accounts of the species.

FEMORAL PORES.—The femoral pores vary geographically within each species, and there are average differences between the species (Table III). The range of variation in femoral pores is from 28 to 52. These counts are the combined number of both femora; there may be a difference of as many as three pores on the right and left femora. In many groups of lizards the number of femoral pores differ in the sexes; however, despite the statement of Duellman (1954: 11), who indicated marked sexual dimorphism in a small sample of *deppei* in Michoacán, we have found the number of femoral pores in males and females to be essentially the same at any one locality.

PREANAL SCALES.—The number of scales in a row between the ventral plates at the apex of the femora and the granules immediately preceding the anal opening is of taxonomic significance. The importance of this character in distinguishing *Cnemidophorus deppei* from *guttatus* at Tehuantepec, Oaxaca, was first pointed out by Hartweg and Oliver (1937, see Fig. 1). Although the character does not hold up throughout the entire range of the group as it does at Tehuantepec, it does provide a means which in combination with other characters is useful in identifying specimens. The total range of variation in the group is from 4 to 11 scales. In those with four, there is a pair of enlarged scales (counted as one) just anterior to the granules surrounding the anus; anterior to this pair are two single enlarged scales, and anterior to these is a small scale in the apex of the femora. In those lizards with a higher number of preanal scales there may be one, and occasionally two, more enlarged scales; the remainder are smaller, and often imbricate, and lie between the large preanals and the apex of the femora.

SUPRAORBITAL SEMICIRCLE SERIES.—This series consists of granules along the median edge of the enlarged supraoculars. In some populations of

Cnemidophorus the series are completely developed; that is, they extend the length of the median edge of the supraoculars and meet the row of granules between the superciliaries and the supraoculars, so as to surround the latter. Such is the case with *Cnemidophorus calidipes* in the *sacki* group (Duellman, 1955). The development of this series in the *deppei* group varies greatly. Usually the series extend no farther anteriorly than the posterior edge of the frontal. Three populations are striking exceptions; most *guttatus* in southern Veracruz, *duodecemlineatus* in Nayarit, and *deppei* from Las Lisas, Guatemala, have the series extending to the anterior edge of the second supraocular.

OTHER HEAD SCALES.—Several other aspects of scutellation were investigated; these include contact of the upper preocular with the upper labials causing a separation of the lower or posterior preocular from the loreal, contact of the postnasal with the third upper labial, number of rows of granules between the superciliaries and supraoculars, development of the row of granules between the lower labials and the sublabials, and relative proportions of length and width of the prefrontals and frontals. The limited investigation of these characters showed that they were highly variable within each population examined; consequently, they are considered to have no taxonomic value in the *deppei* group. One form, *C. deppei cozumelus*, has one to four accessory scutes between the parietals and interparietals; all but one specimen of this race has these scales, which are not present in other populations.

COLORATION

The use of color and pattern in the *deppei* group is of immense importance, particularly in the delimitation of the subspecies. In order to describe completely the color pattern in any one form, an ontogenetic series is mandatory, for there are often drastic changes in pattern from young to old individuals. Many large series of specimens, particularly those collected in the dry season, do not contain adults. Certain color characters, such as the color of the middorsal stripe, throat, and lateral fields often change in preservative, so that strikingly different individuals in life may have much the same color in preservative. Consequently, whenever possible we have supplemented the descriptions of preserved specimens with field notes on living individuals.

DORSAL PATTERN.—Juvenile members of the *deppei* group are black with a variable number of longitudinal light stripes (*guttatus guttatus* is an exception). From this basic pattern there may be a fusion of stripes, loss of stripes, addition of stripes, or replacement of stripes by spots. In order to describe accurately the dorsal color pattern and its ontogenetic manifestations, a terminology for the stripes and intervening fields is necessary (Fig. 1).

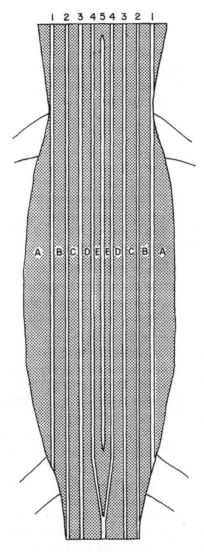

FIG. 1. Teminology of dorsal color pattern in *Cnemidophorus, deppei* group. Stripes: 1, lateral; 2, lower dorsolateral; 3, upper dorsolateral; 4, paravertebral; 5, vertebral. Dark fields: A, flank; B, lateral; C, lower dorsolateral; D, upper dorsolateral; E, paravertebral.

1. Lateral: a pair of stripes beginning immediately behind the eye, passing through the upper edge of the ear opening, just above the insertion of the front limb, and continuing onto the anterodorsal surface of the thigh.

2. Lower Dorsolateral: a pair of stripes originating on the superciliaries above the eye and continuing onto the dorsolateral surface of the tail.

3. Upper Dorsolateral: A pair of stripes arising on the level of the lateral edges of the parietal scales and extending onto the dorsal surface of the tail; absent in *guttatus*.

4. Paravertebral: a pair of stripes arising on the level of the median edges of the parietal scales and usually meeting on the rump to continue as a single stripe onto the tail.

5. Vertebral: a single stripe or pair of stripes lying between the paravertebral stripes, usually not reaching the head, and terminating on the rump. If the vertebral is paired, the stripes meet anteriorly and posteriorly. In some forms the vertebral is absent; frequently in adults it is fused with the paravertebrals.

In adults of many forms the vertebrals and paravertebrals fuse to form one broad middorsal stripe; in order not to confuse this with the original vertebral stripe present in the juvenile, we call the stripe formed by the fusion of the vertebrals and paravertebrals the middorsal stripe. Occasionally in *duodecemlineatus* and *guttatus* an indistinct stripe or row of spots is present between the vertebral stripes; we have not given this a name.

The terminology for the dark fields follows that of the stripes. The fields are, in order, ventral to dorsal: flank, lateral, lower dorsolateral, upper dorsolateral, paravertebral, and vertebral (Fig. 1). In forms in which there is a single vertebral stripe, or in which the vertebral stripe is absent, there is, of course, no vertebral dark field. Likewise, in *guttatus*, which has no upper dorsolateral stripe, there is only one dorsolateral field on each side. Except on the flanks, spots do not form in the dark fields. Spots do form in the stripes, and in some forms these expand to form light bars connecting adjacent stripes.

Both interspecific and intraspecific variation exists in the relative width of the dark fields. This variation may be expressed in the number of granules in a particular field at midbody (Table V).

VENTRAL COLORATION.—The ventral coloration in the adult males of the three species is decidedly different. In *deppei* (subspecies *cozumelus* is an exception), the throat, belly, and ventral surfaces of the hind limbs are black. Both *lineatissimus* and *guttatus* have blue bellies; the former has a pinkish blue chin, and the latter has an orange or buff chin. These colors fade rapidly in preservative, so they have limited use in preserved specimens.

ONTOGENETIC CHANGE IN COLOR PATTERN.—As stated previously, all of the forms in the *deppi* group undergo some metamorphosis of color pattern from juvenile to adult. Usually this involves a fusion of the vertebral and paravertebral stripes and/or a fragmentation of some or all of the stripes into spots; in some it involves the formation of vertical bars on the flanks. This concurrent change of several aspects of the color pattern is difficult to describe. The brief descriptions of ontogenetic change in color pattern accompanying the accounts of the subspecies should be used in conjunction with the diagrams of color pattern metamorphosis (Figs. 11–16). These illustrations are diagrams of three stages of ontogenetic change—juvenile, subadult, and adult male.

SEXUAL DIMORPHISM

We have had for study adequate samples of males and females of each form. No significant sexual dimorphism was found in characters of scutellation; for example, see analysis of sexual differences in *C. deppei* from the Tepalcatepec Valley in Michoacán (Table IV). Males attain a greater size than females (Table I). Apparently correlated with their larger size is their more advanced metamorphosis of color pattern. This is particularly true in *lineatissimus* and *deppei deppei*. In these forms the largest females have color patterns like subadult males. Females have immaculate creamy white ventral surfaces, whereas males have either black ventral surfaces (*deppei*), or blue bellies and either pink (*lineatissimus*) or orange or buff throats (*guttatus*). The ventral coloration and the nature of the femoral pores (males have noticeably larger femoral pores than females) are accurate means of determining the sex of specimens.

TABLE IV

SEXUAL DIMORPHISM IN SCUTELLATION IN *Cnemidophorus deppei* FROM APATZINGAN, MICHOACAN

Character	29 Males		34 Females	
	Range	Mean ± SE	Range	Mean ± SE
Dorsal granules	92–100	97.4 ± 0.62	92–107	98.1 ± 0.52
Femoral pores	33–39	36.1 ± 0.38	32–39	34.9 ± 0.28
Preanal scales	5–8	6.6	5–8	6.4

SYSTEMATICS OF THE *DEPPEI* GROUP

DEFINITION

The *deppei* group of the genus *Cnemidophorus* is composed of three species (*deppei, guttatus,* and *lineatissimus*) with a combined range extending from Veracruz and Nayarit, México, to Costa Rica. The species in this group are alike in, but differ from other species in the genus by, the following combination of characters: eight longitudinal rows of enlarged ventral scales, enlarged mesoptychials, two frontoparietals, three parietals, three supraoculars, granular postantebrachials, no anal spurs, and stripes at least on sides of the young.

HISTORICAL REVIEW

C. E. Burt (1931) summarized the existing literature on this group. He recognized three forms —*deppei deppei, deppei cozumelus,* and *guttatus.* Hartweg and Oliver (1937) provided a workable diagnosis for *guttatus* and *deppei.* They revived Cope's (1877) name *lineatissimus* for the population of *deppei* in Colima and Cope's (1877) name *immutabilis* for the Pacific Coast population of *guttatus.* Smith (1939) described *C. deppei oligoporus* from southern Veracruz. Lewis (1956) described *C. guttatus duodecemlineatus* from Nayarit, which Zweifel (1959) placed as a subspecies of *deppei.* Thus, at the present time two species with a total of seven races are recognized in the *deppei* group. Our study shows the existence of three species with a total of ten races, four of which are described in this paper.

KEYS TO THE SPECIES AND SUBSPECIES
OF THE *DEPPEI* GROUP

The great differences in color pattern at various stages of life make the construction of a key for the identification of these lizards especially difficult. This situation has been partially remedied by the presentation of two keys—one to adults and the other to juveniles. Because of the difficulty in identifying subadult individuals, it might be suggested that collectors either disregard these in the field or collect them alive and raise them to maturity before attempting to identify them by the use of the following keys. Knowledge of the provenance of the specimen is of importance in identifying certain juveniles, and to a lesser extent, some adults; therefore, brief statements of range have been included in the keys. The key to the juveniles is designed to identify hatchlings and small juveniles. Specimens with any spots in the stripes or snout-vent lengths of more than about 45 mm. cannot be identified in that key. The key to adults is for the identification of adult males. In some instances females can be identified correctly, but because they often have an incomplete metamorphosis of color pattern, and thus resemble immature males, many adult females do not possess the diagnostic characters used in the key.

KEY TO THE IDENTIFICATION OF JUVENILES

1. Dorsum brown; only a lateral light stripe present; more than 180 dorsal granules (Veracruz) ....*guttatus guttatus*

   Dorsum black; seven to eleven longitudinal light stripes; dorsal granules variable.. 2

2. Dorsal stripes narrow and wavy; more than 20 granules between paravertebral stripes; 100 to 120 dorsal granules (Yucatán Peninsula) ....*deppei cozumelus*

   Dorsal stripes not wavy; usually less than 20 granules between paravertebral stripes; dorsal granules variable 3

3. Seven stripes (8 if vertebral is bifurcate); no upper dorsolateral stripe; more than 140 dorsal granules 4

   Eight or more stripes; if 8, vertebrals absent; upper dorsolateral stripe present; dorsal granules variable 5

4. Seven stripes, of which vertebral is often bifurcate; 153 to 198 (177) dorsal granules (Pacific Coast of Guerrero and Oaxaca) ....*guttatus immutabilis*

   Seven stripes, of which vertebral is sometimes bifurcate; 142 to 184 (159) dorsal granules (Pacific Coast and Cintalapa Valley, Chiapas) ....*guttatus flavilineatus*

5. Eight or nine stripes (sometimes vertebral, if present, is bifurcate to form ten stripes at midbody); 120 dorsal granules or less and fewer than 16 granules between the enlarged ventrals and lower edge of lateral stripe (except in Chiapas and Guatemala) 9

   Usually ten (sometimes 9 or 11) stripes; more than 108 dorsal granules; more than 15 granules between ventrals and lower edge of lateral stripe (Michoacán to Nayarit) 6

6. Ten or 11 stripes; 125 to 164 dorsal granules (coast of Michoacán, Jalisco, and Nayarit) 7

   Nine or 10 stripes; 110 to 140 dorsal granules (Colima and Tepalcatepec Valley in Michoacán) 8

7. Ten stripes; 126 to 164 (148) dorsal granules; 32 to 48 (38) femoral pores; supraorbital semicircle series not complete (coast of Michoacán)....*lineatissimus lividus*

   Ten or 11 stripes (sometimes extra stripe between vertebrals); 125 to 142 (133) dorsal granules; 28 to 37 (33) femoral pores; supraorbital semicircle series usually complete (Jalisco and Nayarit) ....*lineatissimus duodecemlineatus*

8. Usually 10 stripes; 29 to 39 (33) femoral pores (Colima).... *lineatissimus lineatissimus*

   Usually 9 stripes; 32 to 47 (39) femoral pores (Tepalcatepec Valley in Michoacán)
   .... *lineatissimus exoristus*

9. Usually 8 stripes, of which the lateral is noticeably wider than others; 88 to 120 (99) dorsal granules (Balsas-Tepalcatepec Basin) .... *deppei infernalis*

   Usually 9 (sometimes 8 or 10) stripes, of which the lateral is not much wider than others; 90 to 142 (111) dorsal granules (Guerrero and Veracruz to Costa Rica) .... *deppei deppei*

KEY TO THE IDENTIFICATION OF ADULTS

1. One pair of dorsolateral stripes (or rows of distinct or faint spots); more than 140 dorsal granules; snout-vent length to more than 100 mm. 2

   Two pairs of dorsolateral stripes; less than 165 dorsal granules; snout-vent length usually less than 100 mm. 4

2. Vertebral stripe usually single, broad, and yellow; other stripes persistent; lateral and paravertebral fields darker than dorsolateral field and flank; 142 to 184 (159) dorsal granules (Pacific Coast and Cintalapa Valley, Chiapas).... *guttatus flavilineatus*

   Vertebral stripe usually paired; some or all of stripes replaced by rows of spots; only lateral field darker than others... 3

3. All stripes (lateral sometimes is exception) represented by spots; 184 to 208 (199) dorsal granules; supraorbital semicircle series usually complete (Veracruz) .... *guttatus guttatus*

   Spots formed in all stripes, but most stripes not fragmented into rows of spots; 153 to 198 (177) dorsal granules; supraorbital semicircle series seldom complete (Pacific Coast of Guerrero and Oaxaca) .... *guttatus immutabilis*

4. Spots present in lateral stripe, or lateral stripe fragmented into spots; vertical bars present or not on flanks 6

   Spots not present in lateral stripe; no vertical bars on flanks 5

5. Stripes narrow and wavy; lateral stripe not much wider than others; paravertebrals sometimes indistinct or absent, not fused to form middorsal greenish stripe (Yucatán Peninsula) .... *deppei cozumelus*

   Stripes not wavy; lateral stripe distinctly wider than dorsolaterals; paravertebrals fused with vertebral to form middorsal greenish stripe (Balsas-Tepalcatepec Basin) .... *deppei infernalis*

6. Lateral stripe usually fragmented into row of spots; no distinct vertical bars on flanks; less than 120 dorsal granules (except in Chiapas and Guatemala); paravertebrals distinct or fused with vertebrals (Guerrero and Veracruz to Costa Rica) .... *deppei deppei*

Lateral stripe fragmented or not; distinct vertical bars on flanks, sometimes reaching lower dorsolateral stripe; a yellow middorsal stripe 7

7. Paravertebral stripes fused to form middorsal stripe; lateral stripe usually not fragmented; 108 to 140 (122) dorsal granules (Tepalcatepec Valley in Michoacán) .... *lineatissimus exoristus*

Paravertebral stripes separate from middorsal stripe or absent; lateral stripe fragmented or not 8

8. Flanks and lateral field dark brown or black; upper dorsolateral and paravertebral stripes sometimes indistinct; 125 to 164 (140) dorsal granules 9

Flanks and lateral field not distinctly darker than other fields; 110 to 133 (120) dorsal granules (Colima) .... *lineatissimus lineatissimus*

9. Dorsolateral and paravertebral stripes narrow, but distinct; vertebral stripes fused and bordered by black or dark brown; 126 to 164 (148) dorsal granules; 32 to 48 (38) femoral pores; supraorbital semicircle series not complete (coast of Michoacán) .... *lineatissimus lividus*

Dorsolateral and paravertebral stripes indistinct or absent; middorsal stripe sometimes faint, not bordered by black; 125 to 142 (133) dorsal granules; 28 to 38 (33) femoral pores; supraorbital semicircle series usually complete (Jalisco and Nayarit) .... *lineatissimus duodecemlineatus*

ACCOUNTS OF THE SPECIES AND SUBSPECIES

In this section each species is first treated as a unit in regard to its distribution, diagnostic characters, and geographic variation. The subspecies are treated individually. Each account follows the same pattern; this has been so organized to permit easy comparison of diagnoses, descriptions of ontogenetic change in color pattern, and other characters. It may seem that we have belabored the comparisons of each race with other forms. Granted, there is much repetition; however, the identification of *Cnemidophorus* is difficult. Only by detailed comparisons with all forms with which it might be confused may a specimen or series be identified accurately.

Cnemidophorus deppei Wiegmann

Cnemidophorus deppii Wiegmann, 1834, Herpetologia Mexicana, pp. 27, 28.

DISTRIBUTION.—Lizards belonging to this species range from northern Veracruz and Michoacán southward to Costa Rica, including Isla de Cozumel and Isla Mujeres off the Yucatán Peninsula; they inhabit coastal lowlands and foothills, and the interior valley of the Río Balsas, Río Tepalcatepec, and Río Grijalva in México and the Río Cuilco and Río

Motagua in Guatemala, where they live in open sunny areas to elevations of about 1500 meters (Fig. 2).

DIAGNOSIS.—This is the smallest species in the *deppei* group (largest male with a snout-vent length of 93 mm.) with the lowest number of dorsal granules at midbody (88 to 140). There are 29 to 44 femoral pores and 4 to 9 preanal scales. The supraorbital semicircle series seldom extend anteriorly beyond the posterior edge of the frontal. The dorsal color pattern of juveniles consists of eight or nine longitudinal light stripes on a black ground color. In adults of most populations the paravertebral stripes, and in some the dorsolateral stripes also, fuse to form a broad middorsal light area; the stripes, except the lateral ones, usually are green or greenish yellow. Adult males (except *cozumelus*) have black throats and bellies.

GEOGRAPHIC VARIATION.—Certain geographic trends in scutellation are discernible in this species. There is an increase in the number of dorsal granules at midbody along the Pacific lowlands from Michoacán to Soconusco, Chiapas; there is a similar trend in the number of femoral pores southward to Tonolá, Chiapas, and another in the number of preanal scales (excepting the Balsas Basin) southward to Barra de Cahuacán, Chiapas (Figs. 3 and 4, Table V). Southeastward into Guatemala these trends are reversed. With the exception of the number of dorsal granules in the sample from the Grijalva Valley in Chiapas, the populations in the interior valleys of Chiapas and Guatemala are alike in scutellation. In eastern México there is a sharp break in the number of dorsal granules between the populations in southern Veracruz and that at Ciudad Carmen in Campeche. They are similar in numbers of femoral pores and preanal scales and in coloration. Furthermore, these populations have fewer femoral pores than do those from other parts of the range. A series from Lerdo de Tejada, Veracruz, however, is a striking exception in all characters; in comparison with other samples from southern Veracruz, this series has a higher number of dorsal granules and ten stripes instead of eight. Also, it has a higher number of preanal scales than any other sample of the species.

In coloration there are two distinct populations. One of these on Isla de Cozumel off the northeastern coast of the Peninsula de Yucatán (with disjunct populations on Isla Mujeres and in El Petén, Guatemala) has a broad brown middorsal band and narrow wavy light stripes. The other in the Balsas-Tepalcatepec Basin in western México has a broad persistent lateral light stripe bordered above by deep reddish brown and below by red. In the other populations the lateral stripe fragments into spots in adults. The degree of fusion of the paravertebral stripes and upper dorsolateral stripes varies greatly in these populations.

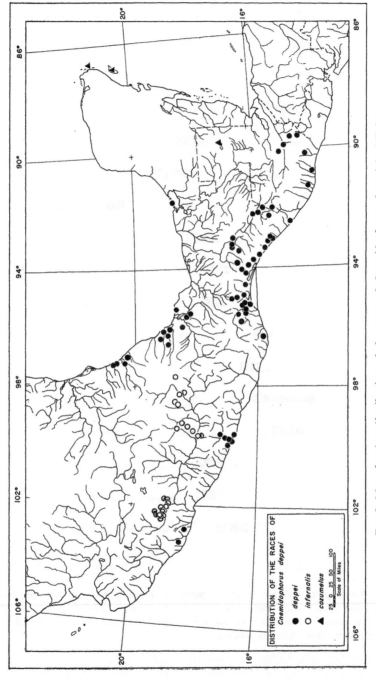

FIG. 2. Map showing distribution of the races of *Cnemidophorus deppei* in México and Guatemala.

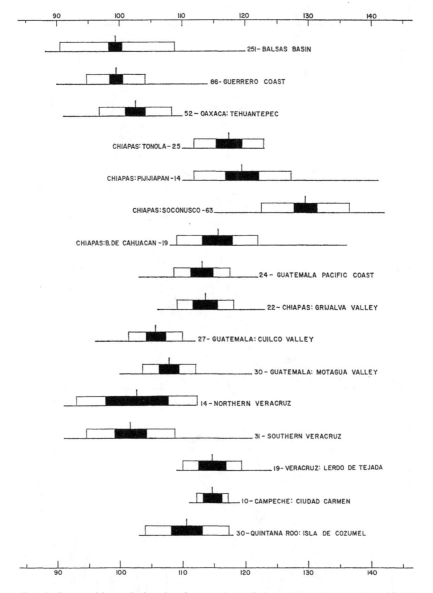

Fig. 3. Geographic variation in the number of dorsal granules in *Cnemidophorus deppei*. Vertical line, mean; horizontal line, observed range; solid box, two standard errors of the mean on either side of the mean; open box, one standard deviation on either side of the mean; number, sample size.

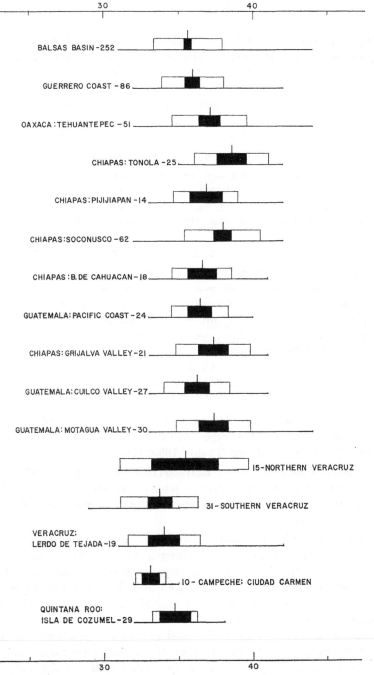

Fig. 4. Geographic variation in the number of femoral pores in *Cnemidophorus deppei*. See Figure 3 for explanation.

The acquisition of additional material from critical areas may show that some of the populations actually represent recognizable races, i.e., the population on the Pacific Coast of Chiapas, or distinct species, i.e., the series from Lerdo de Tejada, Veracruz. On the basis of certain distinguishing characteristics described above only three races are recognized here.

Cnemidophorus deppei deppei Wiegmann

Cnemidophorus deppii Wiegmann, 1834, Herpetologia Mexicana, pp. 27, 28.

Cnemidophorus decemlineatus Hallowell, 1860, Proc. Acad. Nat. Sci. Philadelphia, p. 482 (USNM 6058; type locality "Nicaragua"; collected by C. Wright).

Cnemidophorus lativittis Cope, 1877, Proc. Amer. Philos. Soc., 17: 94 (USNM 30227; type locality "Tuchitan, Tehuantepec" [Juchitán], Oaxaca, México; collected by Francis Sumichrast).

Cnemidophorus deppei deppei, Cope, 1892, Trans. Amer. Philos. Soc., 17: 31.

Cnemidophorus deppei decemlineatus, Cope, 1892, *Ibid.*

Cnemidophorus alfaronis Cope, 1894, Proc. Acad. Nat. Sci. Philadelphia, p. 199 (AMNH 16315; type locality San Mateo, Costa Rica; collected by A. Alfaro).

Cnemidophorus deppii deppii, Burt, 1931, U. S. Natl. Mus. Bull., 154: 56–63 (part). Hartweg and Oliver, 1937, Occ. Papers Mus. Zool. Univ. Michigan, 359: 1–3. Smith and Taylor, 1950, U. S. Natl. Mus. Bull., 199: 178–79.

Cnemidophorus deppei oligoporus Smith, 1939, Field Mus. Nat. Hist., Zool. Ser., 24 (4): 26–7 (CNHM 29145; type locality Pérez, Veracruz, México; collected by Julius Friesser). Smith and Taylor, 1950, U. S. Natl. Mus. Bull., 199: 179.

HOLOTYPE.—Zoologisches Museum, Berlin, No. 882, collected by F. Deppe from "Mexico." The type locality was restricted to Tehuantepec, Oaxaca, México, by Smith and Taylor, 1950: 179.

DIAGNOSIS.—A race of *deppei* characterized by straight stripes, spots in lateral stripes in adults, no accessory scutes between parietals and frontoparietals, and a highly variable number of dorsal granules at midbody (90 to 142, average 110).

DESCRIPTION.—The following description is based on a series of 52 specimens from the vicinity of Tehuantepec, Oaxaca. The largest male has a snout-vent length of 93 mm.; the largest female, 87 mm. Eleven males with snout-vent lengths of more than 80 mm. have tail lengths of 178 to 198 mm. and tail/body ratios of 2.28 to 2.36 (2.33). The smallest juvenile examined has a snout-vent length of 33 mm.

Scutellation: The number of dorsal granules at midbody varies from 91 to 110 (102.5 ± 0.81); the number of femoral pores varies from 32 to 44 (37.1 ± 0.35). There are one or two small scales in the apex of the femora preceding the enlarged preanal scales; the total number of preanal scales varies from 4 to 6 (5.1). In one specimen the supraorbital semicircle series extend anteriorly beyond the posterior edge of the frontal; in all others they extend no farther than the posterior edge of the frontal.

Coloration: Juveniles are black with nine distinct straight creamy white longitudinal stripes, a dark olive gray head, and a blue tail. The limbs are dark brown with irregular cream spots. In subadults the dark fields are dark brown; often the lateral field is reddish brown. The stripes are pale yellow or greenish yellow. In large adults the flanks usually are gray and the middorsum, greenish brown. The spots in the lateral stripe are bluish white. The posterior two thirds of the tail is bluish gray. Adult males have dark bluish gray or black throats and black bellies.

In life the dorsal dark fields are dark brown; the lateral field is reddish chocolate brown. The stripes are pale green, and the tail is pale bluish gray.

ONTOGENETIC CHANGE IN COLOR PATTERN.—The metamorphosis of color pattern in this form involves a replacement of the lateral stripes by spots, usually a splitting of the vertebral stripe, a suffusion of the vertebrals, paravertebrals, and upper dorsolateral stripes into a broad greenish middorsal light area, and sometimes the loss of the lower dorsolateral stripe anteriorly (Fig 11). The vertebral stripe in juveniles usually divides into a pair of narrow vertebral stripes. In older individuals the stripes become indistinct as the vertebral and paravertebral dark fields turn light green. The result of this lightening of the dark fields is a fusion of the vertebral and paravertebral stripes into a broad middorsal light band. In large individuals the upper dorsolateral dark fields turn green, resulting in a fusion of the upper dorsolateral stripes with the light band. In some large individuals the lower dorsolateral stripe becomes indistinct anteriorly. Spots form in the cream lateral stripe, which later changes to bluish white and fragments into a row of spots.

SEXUAL DIMORPHISM.—Males have dark bluish gray or black throats and black bellies and undersurfaces of the limbs. In females the undersurfaces are creamy white. In females the lateral stripes apparently do not break down into spots, nor do the upper dorsolateral stripes become fused with the middorsal light band.

GEOGRAPHIC VARIATION.—More than 400 specimens grouped into fourteen geographic series have been studied in an attempt to analyze variation in this form. Most of the characters of scutellation vary independently from one another and from color differences. Consequently, each of the several characters studied is treated separately.

Dorsal granules (Fig. 3): The lowest number of dorsal granules is found in populations along the Pacific Coast of México from Guerrero to

Oaxaca; the number varies from 90 to 114 (99.4 ± 0.54) in Guerrero and from 91 to 110 (102.5 ± 0.81) in Oaxaca. Southeastward along the Pacific Coast from Tehuantepec there is an increase in the number of dorsal granules at midbody—117.4 ± 1.12 at Tonolá, 119.5 ± 1.33 at Pijijiapan, and 129.5 ± 0.88 at Soconusco, and thence a decrease to 115.7 ± 1.25 at Barra de Cahuacán near Tapachula and 113.1 ± 0.93 on the Pacific lowlands of Guatemala. In the interior valleys of Chiapas and Guatemala are specimens from the Grijalva Valley in Chiapas with 113.6 ± 0.95, from the Cuilco Valley (a tributary of the Río Grijalva in Guatemala) with 105.7 ± 0.83, and from the Motagua Valley in Guatemala with 107.8 ± 0.78. Specimens from the north side of the Isthmus of Tehuantepec in southern Veracruz are similar to those from Tehuantepec in having 101.6 ± 1.27 dorsal granules. A series from Lerdo de Tejada, southeast of Alvardo, is an exception; in this series the number of dorsal granules varies from 109 to 124 (114.7 ± 1.08). A series from Ciudad Carmen, Campeche, has 114.7 ± 0.77. Although similar to the series from Lerdo de Tejada in the number of dorsal granules, the specimens from Ciudad Carmen resemble the other series from southern Veracruz in coloration and numbers of femoral pores and preanal scales. One specimen from Ciudad Carmen is like the others in coloration, but has 159 dorsal granules and 44 femoral pores; this apparently aberrant individual has not been included in this analysis.

Femoral Pores (Fig. 4): Aside from the populations on the Atlantic side of the Isthmus of Tehuantepec, there is little variation in the number of femoral pores. All samples from the Pacific Coast and the interior valleys of Chiapas and Guatemala have averages between 35.9 and 38.6. There is, however, a slight trend from lower to higher numbers of femoral pores from Guerrero to Tonolá and Soconusco, Chiapas; a sample from Guerrero has 32 to 42 (35.9 ± 0.23), as compared with 35 to 42 (38.6 ± 0.49) at Tonolá and 32 to 42 (37.9 ± 0.33) at Soconusco. Specimens from the Atlantic side of the Isthmus of Tehuantepec have slightly fewer femoral pores; the average number for various populations varies from 33.1 to 34.0.

Preanal Scales (Table V): With the exception of the samples from Soconusco, Chiapas, and Lerdo de Tejada, Veracruz, which have two specimens with eight and one with nine, and four with eight and one with nine, respectively, no samples contain specimens with more than seven preanal scales. There is a trend for an increase in preanal scales from Guerrero (average 4.8) to Soconusco, Chiapas (average 6.2), and then a decrease to the Pacific Coast of Guatemala (average 5.2). With the exception of the series from Lerdo de Tejada, Veracruz, which has an average of 7.3, all of the samples from the Atlantic side of the Isthmus of Tehuantepec and from the interior valleys of Chiapas and Guatemala have averages from 5.0 to 5.4.

TABLE V

GEOGRAPHIC VARIATION IN FOUR CHARACTERS IN *Cnemidophorus deppei deppei*
(Observed ranges in parentheses)

Population	N	Preanal Scales	Granules Below Lateral Stripe	Granules in Lateral Dark Field	Granules Between Paravertebral Stripes
Guerrero	86	(4–6) 4.8	(12–17) 15.1	(7–9) 7.8	(4–15) 9.9
Oaxaca: Tehuantepec	52	(4–6) 5.1	(14–17) 15.0	(6–9) 6.9	(10–24) 14.3
Chiapas: Tonolá	25	(5–6) 5.6	(15–19) 16.8	(8–12) 9.9	(2–12) 7.7
Chiapas: Pijijiapan	14	(5–7) 5.8	(15–26) 17.9	(10–11) 10.5	(4–11) 6.8
Chiapas: Soconusco	63	(5–9) 6.2	(18–22) 19.6	(9–12) 10.7	(6–15) 10.5
Chiapas: Barra de Cahuacán	19	(5–7) 6.1	(15–21) 17.9	(9–12) 10.3	(6–13) 9.2
Guatemala: Pacific Coast	24	(4–6) 5.2	(16–18) 16.4	(9–12) 9.9	(9–13) 11.2
Chiapas: Grijalva Valley	22	(4–7) 5.3	(14–18) 16.3	(8–9) 8.5	(10–22) 15.1
Guatemala: Cuilco Valley	27	(5–6) 5.3	(13–16) 14.5	(7–9) 7.8	(10–18) 12.9
Guatemala: Motagua Valley	30	(4–7) 5.4	(13–16) 14.6	(8–9) 8.5	(12–20) 15.7
Northern Veracruz	15	(5–7) 5.7	(11–13) 12.2	(9–12) 10.6	(3–13) 6.5
Southern Veracruz	31	(5–6) 5.3	(11–20) 14.3	(8–11) 9.4	(5–14) 7.2
Veracruz: Lerdo de Tejada	19	(6–9) 7.3	(15–18) 15.8	(7–9) 7.7	(12–20) 16.1
Campeche: Ciudad Carmen	10	(5–6) 5.3	(16–18) 17.1	(9–10) 9.5	(11–14) 12.4

Supraorbital Semicircle Series: In most specimens the supraorbital semicircle series do not extend anteriorly beyond the posterior edge of the frontal; in a few the series terminate between the posterior edge of the frontal and the posterior edge of the second supraocular. The series are complete in 4 per cent of the specimens from Tonolá, Chiapas, 4 per cent from the Grijalva Valley, 7 per cent from the Motagua Valley, and 3 per cent from Soconusco, Chiapas. Of 14 specimens from Las Lisas, Santa Rosa, Guatemala, 10 (71 per cent) have the series complete. In the other samples no specimens have complete semicircle series.

Coloration: Among the series studied the color pattern is highly variable; only one character, the formation of spots in the lateral stripes in adult males, is constant. An analysis of the variation in the relative widths of the dark fields is given in Table IV. Certain characters of coloration appear to be constant in different populations. For example, in specimens from the Pacific Coast of Chiapas and Guatemala the paravertebral stripes seldom, and the upper dorsolateral stripes never, fuse into a broad middorsal band. In some juveniles the vertebral stripe is absent. Specimens from Soconusco, Chiapas, together with some from Pijijiapan and Barra de Cahuacán, have particularly wide and distinct stripes. In two specimens from Tonolá and one from Pijijiapan the paravertebral stripes are fused to form a relatively narrow middorsal stripe. In some specimens from Guerrero the paravertebral stripes are broad and close together. Specimens from Ciudad Carmen, Campeche, and many specimens from Veracruz have the paravertebral stripes separated by a rather broad light brown area without a trace of a vertebral stripe.

The series from Lerdo de Tejada, Veracruz, is unique in possessing ten distinct longitudinal stripes. These are present in juveniles and adult males; in the latter, spots are present in the lateral stripes (Fig. 12), but the stripes are not fragmented into spots as is characteristic of other *deppei*. These color pattern differences together with the greater number of dorsal granules (114.7 $\pm$ 1.08, as compared with 101.6 $\pm$ 1.27) and the greater number of preanal scales (7.3, as compared with 5.3) set this series off from "normal *deppei*" as it occurs in southern Veracruz.

Ignoring the Lerdo de Tejada population for a moment, we are inclined to refer all of the populations discussed above to one race, *Cnemidophorus d. deppei*. Granted, some populations are endowed with characteristics distinguishing them from adjacent populations; however, these characters usually are present in other disjunct populations, i.e., the broad stripes in Chiapas and Guerrero. A distinct geographic break is present in each of two characters. Between Tehuantepec, Oaxaca, and Tonolá, Chiapas (an airline distance of about 150 kilometers), there is a decided discrepancy in the number of dorsal granules at midbody; at Tehuantepec the number varies from 91 to 110 (102.5 $\pm$ 0.81), and at Tonolá, from 110 to 123 (117.4 $\pm$ 1.12). The other break is in the number of femoral pores between the populations on the Atlantic side of the Isthmus of Tehuantepec and those on the Pacific side; on the Atlantic side the number varies from 29 to 35 (33.7 $\pm$ 0.38) in southern Veracruz and from 32 to 35 (33.1

± 0.31) at Ciudad Carmen, as compared with 32 to 44 (37.1 ± 0.35) at Tehuantepec. The taxonomic significance of these breaks is minimized by the independent variation of other characters. Although there is a definite difference in the number of femoral pores, and a minor difference in color pattern between the populations on the Atlantic side of the Isthmus of Tehuantepec and that at Tehuantepec on the Pacific side, there is practically no difference in the number of dorsal granules, nor in the number of preanal scales. Furthermore, low numbers of femoral pores occur in the population in Guerrero. With respect to the difference exhibited by the population on the coast of Chiapas, it should be pointed out that although there is an increase in the number of preanal scales and femoral pores together with the increase in the number of dorsal granules at midbody, the high counts are reached at various places. Furthermore, southeastward there is a decrease in the number of scales. Thus, on the basis of our present knowledge of variation in this form we believe that the independence of variability in the characters studied does not show distinct breaks that are significant on the subspecific level.

Smith (1939) described the eight-striped population from Veracruz as *C. d. oligoporus*, and diagnosed it as having a significantly lower number of femoral pores. Examination of additional material substantiates the lower average number of femoral pores in this population. However, if on the basis of the average difference in the number of femoral pores this population is recognized as a distinct taxonomic unit, the population on the coast of Chiapas (characterized by its higher number of dorsal granules, more femoral pores and more preanal scales) likewise should be named. The same criteria would necessitate the recognition of the population at Ciudad Carmen as distinct from *oligoporus*, for the former has a significantly higher number of dorsal granules. The minor differences between the populations in the interior valleys of Chiapas and Guatemala and those from the coastal lowlands would necessitate giving a name to those in the interior valleys. Therefore, if populations were named on the basis of these minor differences, *deppei deppei* as recognized here would consist of five races. Furthermore, such criteria would not be consistent with those used here for the other species in the *deppei* group. Until additional data concerning variation, habitat, and distribution are available for the southern populations, we feel that a conservative approach is best.

The series from Lerdo de Tejada presents another problem. To the north at Veracruz and to the south at Cuatotolapam and Pérez are found specimens with eight stripes and low numbers of dorsal granules and preanal scales. No individuals of this kind were found with the specimens collected on the leeward side of the coastal dunes near Lerdo de Tejada.

Perhaps intensive collecting along the beaches in Veracruz will provide evidence that this ten-striped form with a high number of granules has an extensive range in that habitat and at least in some areas occurs sympatrically with "normal *deppei*," in which case it should be treated as a distinct species.

The specimens from Ciudad Carmen, Campeche, and some of those from southern Veracruz have relatively broad middorsal brown bands between the paravertebral stripes and no trace of a vertebral stripe. In this respect they resemble *C. d. cozumelus*; however, the latter has wavy stripes and accessory scutes between the parietals and frontoparietals, characters not present in specimens from Ciudad Carmen and Veracruz. The reddish brown lateral field characteristic of the population in the Balsas Basin is found to some extent in specimens from Guerrero and Oaxaca, and even in the ten-striped population at Lerdo de Tejada, Veracruz. However, all specimens from the Pacific Coast of México have the lateral stripes fragmented into spots in adults; this is never so in specimens from the Balsas Basin. No intergrades between *deppei* and *infernalis* are known; they might be sought for in the lower reaches of Balsas Valley.

No attempt has been made to study *deppei* in the southern part of its range, from Honduras to Costa Rica. Possibly the acquisition of sufficient material together with a knowledge of the animals in life from the southern part of the range will elucidate some of the problems of variation that we have encountered.

Comparison With Other Forms.—The variability of this form makes comparisons difficult. However, a few constant characters serve to distinguish it from the other races. No other race of *deppei* has spots in, or in place of, the lateral stripe. In *infernalis* the lateral stripe is broad and distinct; in *cozumelus* the stripes are wavy. From *guttatus,* with which *deppei* occurs throughout most of its range, *deppei* may be distinguished by its lower number of dorsal granules at midbody (usually [99 per cent] less than 140, as compared with more than 140 in *guttatus*) and the different ventral coloration of adult males. In *guttatus* the throat is orange or buff, and the belly is blue, whereas in *deppei* the belly is black, and the throat is dark bluish gray or black.

Ecological Notes.—*Cnemidophorus deppei deppei* most frequently is found in open areas, often where the soil is sandy and there is a sparse growth of grass. Often it is common along river flood plains and other riparian situations in subhumid environments. Throughout most of its range (except the interior valleys of Chiapas and Guatemala, Ciudad Carmen, and the Pacific Coast from Soconusco southward) it occurs with

C. guttatus, and throughout its entire range, with *Ameiva undulata*. Both of these species appear to prefer shaded areas.

DISTRIBUTION.—Southern Veracruz and southern Michoacán southeastward at elevations less than 1000 meters to the Isthmus of Tehuantepec and Isla de Carmen and thence southeastward through the interior valleys of Chiapas and Guatemala (Grijalva and Motagua valleys) and on the coast of Chiapas southward to Costa Rica. Locality records (1000 specimens) follow.

Guatemala: *Escuintla*: Tiquisate, UMMZ 98189 (2). *Huehuetenango*: Finca Canibal, Río Cuilco, UMMZ 98196 (27), 98197 (5). *Jutiapa*: Finca La Trindad, UMMZ 107481–92 (63); Hacienda Mongoy, UMMZ 106987–90 (32); Santa Catarina Mitla, UMMZ 106991 (10). *Progreso*: El Rancho, UMMZ 106981–6 (52); Finca Los Leones, UMMZ 106992 (10). *Santa Rosa*: Las Lisas, UMMZ 107479 (14).

México: *Campeche*: Ciudad Carmen, UIMNH 35976–86. *Chiapas*: Arriaga, UMMZ 94838–9, 117495; Barra de Cahuacán, UMMZ 88371 (19); 4 km. S of Chiapa de Corzo, UMMZ 94843–72; Colonia Soconusco, UMMZ 87044–8; crest above Arriaga, UMMZ 94874; Cruz de Piedra, UMMZ 87049–66; El Ocotal, UMMZ 109569 (3); Escuintla, UMMZ 87005–7; between Escuintla and Acacoyagua, UMMZ 87008–43; Finca San Bartolo, UIMNH 39254; Hacienda Monserrate, UMMZ 102223 (3); Mapastepec, UMMZ 105398–9 (8); Mazapa, UMMZ 94826–7; Pijijiapan, UMMZ 88372 (13), 88373; Puerto Madero, UKMNH 43950–44003; Río San Gregorio, UMMZ 109568, 109570; 5 km. S of San Antonio, UMMZ 114787; Tapachula, UIMNH 36124–7; Tonolá, UMMZ 88374–7 (25); 10 km. NW of Tonolá, UKMNH 43906–13, 43915–21, 43923–5, 43928–49; Tuxtla Gutierrez, UMMZ 94873; 10 km. W of Tuxtla Gutierrez, UMMZ 94840–2; Villa Flores, UMMZ 105393 (12), 105394–7. *Guerrero*: Acapulco, UIMNH 36297, 38090, UMMZ 112947; 5 km. N of Acapulco, UIMNH 36280–2; 32 km. SE of Acapulco, UIMNH 39283–90; Coyuca, UIMNH 36283–8, UMMZ 85411 (7), 105614 (3); 8 km. E of Coyuca, UIMNH 36276–9; El Limón, UIMNH 36023; El Treinta, UIMNH 36034–5, UMMZ 119142; Organos, UIMNH 36019–20; Pie de la Cuesta, UMMZ 104451; Tierra Colorado, UIMNH 36031, 36242–74, 36290–6, UMMZ 80948 (2); Xaltinanguis, UIMNH 36021–2, 36032–3. *Michoacán*: Salitre de Estopila, UMMZ 104530; San Pedro Naranjestila, UMMZ 104531. *Oaxaca*: Cajón de Piedras, UIMNH 36377–9; 10 km. NW of Camarón, UMMZ 114785 (14); Cerro de Huamelula, UIMNH 36367–9; Cerro Quiengola, UIMNH 28058; Chacalapa, UKMNH 38255–9; Ejutla, UMMZ 105410; 3 km. N of Matías Romero, UMMZ 114784; 10 km. W of Nejapa, UKMNH 44045–81; 13 km. E of Pachutla, UIMNH 8482–4; Portillo Las Vacas, UIMNH 36370–6; Río Tequisistlán, UIMNH 39359–81; Salina Cruz, UMMZ 81866–7 (4), 119143 (7); San Geronimo, UIMNH 8485; San Juanico, UMMZ 113824; Tapanatepec, UMMZ 84492 (3); Tehuantepec, UIMNH 8487–98, 36380–36450, UMMZ 81859–65 (19), 81868–73 (11); 5 km. W of Tehuantepec, UKMNH 33720, 37859–73, UMMZ 112948–9; 8 km. W of Tehuantepec, UKMNH 44004–43; 10 km. W of Tehuantepec, UKMNH 39720–1; 12 km. W of Tehuantepec, UMMZ 114786; 16 km. W of Tehuantepec, UKMNH 33715, 33717, 33719, 33721; 32 km. W of Tehuantepec, UMMZ 112663; 18 km. W of Tequisistlan, UMMZ 117494 (5); Totolapam, UIMNH 35995–36005; Unión Hidalgo, UMMZ 113828; Ventosa, UIMNH 26024–5. *Veracruz*: Alvarado, UIMNH 35315; Cempoala, UIMNH 26029–33; Cuatotolapam, UMMZ 41502–4; La Laja, UIMNH 3961–2; 5 km. NW of Lerdo de Tejada, UMMZ 114788 (19);

Matacabresto, UIMNH 36451–3, UMMZ 88650 (11); 9 km. NW of Nautla, UKMNH 24095–9; Otoba, CNHM 1313 (2); Pérez, CNHM 1683 (2); Puente Colorado, UMMZ 89322 (7); Río Blanco, 20 km. WNW of Piedras Negras, UKMNH 23257–61, 24433; Rodriquez Clara, UIMNH 35993; 3 km. S of Tecolutla, UIMNH 3903–8; 16 km. S of Tecolutla, UIMNH 3938–44; Tierra Colorado, UIMNH 35993, UMMZ 95107; Veracruz, CNHM 1343 (3); 50 km. W of Veracruz, UMMZ 99914.

Cnemidophorus deppei infernalis, new subspecies[2]

Cnemidophorus deppei, Gadow, 1906, Proc. Zool. Soc. London, pp. 309–16 (part).

Cnemidophorus deppii deppii, Burt, 1931, U. S. Natl. Mus. Bull., 154: 56–62 (part).

Cnemidophorus deppii lineatissimus, Schmidt and Shannon, 1947, Fieldiana-Zool., 31 (9) ; 75. Smith and Taylor, 1950, U. S. Natl. Mus. Bull., 199: 179 (part). Davis and Smith, 1953, Herpetologica, 9: 106.

Cnemidophorus deppei lineatissimus, Duellman, 1954, Occ. Papers Mus. Zool. Univ. Michigan, 560: 11; 1955, *Ibid,* 574: 6.

HOLOTYPE.—University of Michigan Museum of Zoology No. 114783, from Mexcala, Guerrero, México (on Río Balsas at an elevation of 370 meters), collected by William E. Duellman, June 21, 1956. Original number WED 9361 (Pl. I).

PARATOPOTYPES.—UIMNH 36030, UMMZ 119298–119301.

DIAGNOSIS.—A race of *deppei* characterized by a low number of dorsal granules at midbody (average about 99), straight stripes of which the lateral is broad, cream, and persistent throughout life, a broad reddish brown lateral field, bright reddish brown flanks, paravertebral stripes fused into a broad greenish middorsal stripe in adults, and no accessory scutes between the parietals and frontoparietals.

DESCRIPTION OF HOLOTYPE.—An adult male with a snout-vent length of 78 mm., a tail length (complete) of 172 mm., and a tail/body ratio of 2.20. Scutellation is typical of the *deppei* group—three supraoculars, enlarged mesoptychials, and granular postantebrachials. The supraorbital semicircle series extend anteriorly only to about the middle of the frontoparietals. There are 100 dorsal granules at midbody, 36 femoral pores, and four preanals.

The top of the head, rostral, nasals, and upper parts of the postnasals and loreals are light olive brown. The sides of the head, upper labials, and mental are bluish gray mottled with cream. The upper surfaces of the limbs and tail are dark brown. There is a row of indistinct cream spots on each forelimb. The lower labials, throat, belly, preanal region, and

___

[2] Latin, *infernalis,* belonging to the lower regions, in this case the Balsas Basin, the classical "tierra caliente" of México.

ventral surfaces of the thighs are black. The undersides of the forelimbs, shank, and tail are cream. The paravertebral stripes are fused into a broad middorsal greenish gray stripe. The dorsolateral stripes are distinct and greenish gray. The lateral stripes are broad, creamy white, and separated from the enlarged ventrals by 14 granules. The flanks just below the lateral stripe are light reddish brown fading to a grayish tan ventrally. The lateral fields are dark reddish brown and 11 granules in width at midbody. The other dark fields are dull brown.

In life the lateral stripe was bright creamy white; the other stripes were pale green. The lateral dark fields were rich reddish brown; the upper parts of the flanks were a brighter reddish brown, and the other dark fields were dark brown. The ventral surfaces were deep bluish black.

DESCRIPTION OF THE SUBSPECIES.—The following description is based on 252 specimens from the Balsas-Tepalcatepec Basin. The largest male has a snout-vent length of 84 mm.; the largest female, 75 mm. Twenty males with snout-vent lengths of more than 70 mm. have tail lengths of 172 to 183 mm. and tail/body ratios of 2.20 to 2.52 (2.36). The smallest juvenile examined has a snout-vent length of 30 mm.

Scutellation: The number of dorsal granules at midbody varies from 88 to 120 (99.3 $\pm$ 0.56); only 16 specimens (6.3 per cent) have more than 110 granules. The number of femoral pores varies from 31 to 44 (35.6 $\pm$ 0.14). Usually there are one or two small scales in the apex of the femora; the number of preanal scales varies from 4 to 8 (6.3). The supraorbital semicircle series never extend beyond the posterior edge of the frontal; in most specimens they terminate near the middle of the frontoparietals.

Coloration: Juveniles are black with grayish brown heads, dark brown limbs, and pale blue tails. There are eight longitudinal light stripes, of which the laterals are distinctly wider than the others. An occasional juvenile has an indistinct vertebral line. In adults and subadults the para-vertebral and dorsolateral stripes are pale green, and the lateral stripes are cream. The dark fields, except the lateral which is reddish brown, are dark brown. The flanks immediately below the lateral stripe are light reddish brown changing to grayish tan ventrally. The head is olive brown, and the tail is dark brown. In large specimens the paravertebral stripes are fused to form a broad middorsal green stripe. There are 12 to 16 (13.4) granules between the enlarged ventrals and the lower edge of the lateral stripe and 9 to 12 (10.7) granules in the lateral dark field.

ONTOGENETIC CHANGE IN COLOR PATTERN.—The principal change in color pattern in this form consists of the fusion of the paravertebral stripes into a broad middorsal stripe (Fig. 11). Juveniles usually have eight creamy white longitudinal stripes; some have a faint vertebral stripe. In subadults a vertebral stripe usually is present. In adults the paravertebrals fuse with the vertebral stripe. This fusion begins anteriorly; only in large males are the stripes fused for their entire length. In adults and subadults the paravertebrals and dorsolaterals are light green. Spots do not form in any of the stripes. In large males the dorsolateral stripes may be indistinct, especially posteriorly, but the lateral stripe persists as a broad cream stripe. The dark fields change from black to brown. The lateral fields are reddish brown in adults and subadults. In large males the lower flanks may be grayish brown or grayish green in contrast to the bright reddish brown of the upper flanks.

SEXUAL DIMORPHISM.—Males attain a larger size than females, have a more advanced color pattern in that the paravertebral stripes are fused for their entire lengths, and have black ventral surfaces. The black of the ventral surfaces develops first anteriorly. In subadult males only the throat and chest are black; the belly is blue or bluish gray. In some large males there are small bluish white spots along the lateral edges of the black belly. Females have an immaculate creamy white belly.

GEOGRAPHIC VARIATION.—There are some minor differences in scutellation between the series of 234 specimens from the Tepalcatepec Valley in Michoacán and 18 specimens from the upper Balsas Basin in Guerrero, Morelos, and Puebla. Possibly these differences are a reflection of the discrepancy in sample size. Those from Michoacán have 91 to 120 (101.4 ± 0.58) dorsal granules, as compared with 88 to 110 (96.1 ± 1.73) in the upper Balsas Basin, and 31 to 43 (35.6 ± 0.14) femoral pores, as compared with 31 to 44 (35.4 ± 0.82). Specimens from Michoacán have more preanal scales (6.6) than those from the upper Balsas Basin (5.4).

COMPARISON WITH OTHER FORMS.—From the nominal race, *infernalis* may be distinguished by the persistent cream lateral stripe in adults. In *deppei* the stripe is bluish white; spots form in the stripe which often fragments into a row of spots. In some *deppei* the upper dorsolateral stripes are fused with the paravertebrals; the dorsolateral stripes are never fused in *infernalis*. Some *deppei* have a reddish brown lateral field, but this is never as red as in *infernalis*; furthermore, *deppei* does not have bright reddish brown flanks. Only in specimens of *deppei* from the coast of Chiapas are the lateral fields as broad as in *infernalis* (Table V). From *cozumelus*, *infernalis* differs in having straight stripes, fused paravertebral stripes, and no accessory scutes between the parietals and frontopareitals. From *C. lineatissimus exoristus*, which occurs sympatrically with *infernalis* in the Tepalcatepec Valley, *infernalis* may be distinguished by the absence of lateral spots and vertical bars and by its black belly and throat instead of a blue belly and pinkish blue throat as in *exoristus*. Juveniles of the two forms may be distinguished by the number of granules between the ventrals and the lateral stripe; there are 12 to 16 in *infernalis* and 18 to 23 in *exoristus*.

ECOLOGICAL NOTES.—In the Balsas-Tepalcatepec Basin this lizard is encountered in riparian situations, in grassy areas in barrancas, and in open scrub forest where it may be associated with *Cnemidophorus calidipes* and *C. sacki* (*auctorum*).

DISTRIBUTION.—The range of this race encompasses the arid Balsas Valley and Tepalcatepec Valley; in this interior basin it occurs at elevations from 200 to 1500 meters. Locality records (287 specimens) follow.

México: *Guerrero*: Mexcala, UIMNH 36030, 36101, UMMZ 114783, 119298–301; 3 km. S of Mexcala, UMMZ 114782, 119296–7; 14 km. N of Zumpango del Río, UMMZ 104452. *Michoacán*: Acahuato, CNHM 38976; Apatzingán, CNHM 36976–82, 38974 (23), 38975 (42), UIMNH 36454–520, 36676–8, UMMZ 85414 (4), USNM 135701–6, 135708–45, 135822–4, 135967; 5 km. W of Apatzingán, UKMNH 29187–8, 29190, 29193–5, 29197, 29200, 29204–7, 29289; 6 km. E of Apatzingán, UMMZ 114776 (2); 9 km. E of Apatzingán, UMMZ 112656 (4); 12 km. E of Apatzingán, UMMZ 112652, 112658, 114775 (3), 114777; 15 km. E of Apatzingán, UMMZ 114770, 114778; 13 km. S of Apatzingán, UKMNH 29192, 29199, 29684, 29686; 4 km. N of Capirio, UMMZ 114771; El Capire, CNHM 36985–7; Hacienda California, CNHM 36983–4; Hacienda El Sabino, UIMNH 36100; Jazmin, UMMZ 114773; La Playa, UMMZ 104738, 105241 (2); between La Playa and Jorullo, UMMZ 104739, 104749 (2); 12 km. S of Lombardia, UKMNH 29198, 29201–3; Puerto Crucita, UIMNH 35994; Río Marquez, 10 km. S of Lombardia, UMMZ 112655, 112657, 112661 (3), 114769; Río Marquez, 13 km. SE of Nueva Italia, UMMZ 114774 (9), 119351; south of Tancítaro, CNHM 36975; Volcán Jorullo, UMMZ 108009 (3). *Morelos*: 9 km. W of Alpuyeca, UIMNH 26027; Cuernavaca, UIMNH 35975; Lago Tequesquitengo, UMMZ 114767; 19 km. S of Puente de Ixtla, UIMNH 36029. *Puebla*: 11 km. SW of Izúcar de Matamoros, UKMNH 39710; Río Atoyac, 10 km. N of Tezuitzingo, UIMNH 36028; 3 km. NW of Tezuitzingo, UMMZ 114768 (2); Tilapa, UKMNH 38269; Zapotitlan, UIMNH 36036–7.

Cnemidophorus deppei cozumelus Gadow

Cnemidophorus deppei cozumela Gadow, 1906, Proc. Zool. Soc. London, p. 316.

Cnemidophorus deppii cozumelus, Burt, 1931, U. S. Natl. Mus. Bull., 154: 63–5. Smith and Taylor, 1950, *Ibid.*, 199: 179.

SYNTYPES.—British Museum (Natural History) Nos. 1886.4.15.17–20 (re-registered: 1951.1.8.24–27) from Isla de Cozumel, Quintana Roo, México; collected by Gaumer.

DIAGNOSIS.—A race of *deppei* with narrow wavy longitudinal stripes, a broad middorsal brown band, an average of about 110 dorsal granules at midbody, and usually one to three accessory scutes between the parietals and frontoparietals.

DESCRIPTION.—The following description is based on 30 specimens from Isla de Cozumel. The two largest specimens are females with snout-vent lengths of 83 mm.; the largest male has a snout-vent length of 77 mm. Five males with snout-vent lengths of more than 60 mm. have tail lengths of 143 to 167 mm. and tail/body ratios of 2.17 to 2.37 (2.26). The smallest specimen examined has a snout-vent length of 48 mm.

Scutellation: The dorsal granules vary in number at midbody from 103 to 118 (110.6 ± 1.23). The femoral pores vary from 32 to 38 (34.7 ± 0.27). There are 4 to 6 (5.0) preanal scales. The supraorbital semicircle series do not extend anteriorly beyond the posterior edge of the frontal; in some individuals the series terminate near the middle of the frontoparietals. Lying between the parietals and the frontoparietals are one to three accessory scales. In each of 12 individuals there is one scale; in 7 there are two, in 10 there are three, and in one there is none.

Coloration: We have not seen living individuals of this form; therefore, the color description given below is not as nearly complete as for the other forms discussed in this paper. The top of the head is olive brown, and the sides of the head are bluish white. The upper surfaces of the tail and limbs are olive brown and unspotted. The throat is white, sometimes tinged with blue; the belly is white or bluish white, and the underside of the tail is cream sometimes tinged with blue. There are eight narrow longitudinal stripes (no vertebrals); these stripes are wavy. In large specimens the dorsolaterals are indistinct anteriorly and posteriorly, and the paravertebrals may have disappeared. The flanks below the lateral stripe are bluish gray; sometimes there is an indistinct lower lateral stripe extending from the axilla to the groin. The lateral and dorsolateral fields are dark brown; the lateral one is slightly wider than the others. The middorsal field is light brown to light olive green; there are 21 to 31 granules between the paravertebral stripes. There are 17 to 20 granules between the enlarged ventrals and the lower edge of the lateral stripe and 7 to 9 granules in the lateral field.

ONTOGENETIC CHANGE IN COLOR PATTERN.—The present series does not include hatchlings; consequently, a complete picture of the ontogenetic change in color pattern cannot be presented. Assuming that hatchlings are black as in the other races of *deppei*, we have so depicted it in Figure 12. Possibly a vertebral stripe or pair of vertebrals are present in hatchlings and subsequently are lost. Furthermore, our series contains no males with black bellies, a characteristic of adult males of other races of *deppei*. We do not know whether this character develops in *cozumelus*. Actually little metamorphosis of color pattern is evident in our series. The upper dorsolateral stripes and paravertebral stripes tend to become indistinct anteriorly and posteriorly; in some large specimens the paravertebrals are barely discernible. Two specimens (82 and 67 mm. snout-vent length) have all stripes indistinct; in these there is little difference in color between the lateral and dorsal dark fields. Perhaps this is the coloration typical of adult males. In none is there an indication of the stripes being replaced by spots.

SEXUAL DIMORPHISM.—According to Kathleen Beargie (in *litt.*), who collected this form on Isla Cozumel, the ventral surfaces of both sexes are white. Large individuals of both sexes have pale blue on the posterior part of the bellies.

GEOGRAPHIC VARIATION.—One specimen from Isla Mujeres, Quintana Roo, has 118 dorsal granules at midbody, 5 preanal scales, 33 femoral pores, and two accessory scutes between the parietals and frontoparietals. This specimen has faint stripes and an olive tan ground color. Four specimens from Ramate, El Petén, Guatemala, have 101 to 108 (105.0) dorsal granules, 4 and 5 (4.5) preanal scales, and 35 and 36 (35.2) femoral pores. Of these four specimens, one has 2 accessory scutes, two have 3, and one has 4. In the Guatemalan specimens the stripes are not as wavy as in typical individuals. In other aspects of coloration they show no noticeable differences from specimens from Isla de Cozumel. Data in Tables II and III are based on specimens from Isla de Cozumel only.

COMPARISON WITH OTHER FORMS.—The presence of the accessory scutes between the parietals and frontoparietals, the absence of vertebral stripes, the broad dorsal dark field, and the narrow wavy stripes serve to distinguish *cozumelus* from all other members of the *deppei* group. The only form with which it might be confused is *C. g. guttatus*; the young of this form have only the lateral stripe present and have a tan or olive tan, dorsum. This species has no accessory head scutes and has more than 180 dorsal granules.

ECOLOGICAL NOTES.—No ecological information is available for the specimens from Cozumel and Mujeres islands. The specimens from Ramate, Guatemala, were collected by L. C. Stuart in brushy areas surrounding Lago de Petén, which lies in the savanna country, of southern El Petén.

DISTRIBUTION.—This form is known only from Isla de Cozumel and Isla Mujeres off the northeast coast of the Yucatán Peninsula, and from a disjunct population in the savanna region of El Petén, Guatemala. Possibly future collections from the savanna portions of Campeche and Quintana Roo will reveal additional populations of this lizard. To the south in the Motagua Valley in Guatemala and to the west in southern Veracruz and Campeche occurs *C. deppei deppei*. There is no evidence of intergradation between *deppei* and *cozumelus*. In fact, *cozumelus* logically could be considered a species separate from *deppei*. Locality records (35 specimens) follow.

Guatemala: *El Petén*: Ramate, UMMZ 74979 (4).

México: *Quintana Roo*: Isla de Cozumel, UMMZ 78589 (4), 78590 (2), 78591 (9), 78592 (3), 78593 (12); Isla Mujeres, UMMZ 78588.

Cnemidophorus lineatissimus Cope

Cnemidophorus lineatissimus Cope, 1877, Proc. Amer. Philos. Soc., 17: 94.

DISTRIBUTION.—Lizards of this species range along the Pacific lowlands northwestward from the Río Balsas to central Nayarit and into the Tepalcatepec Valley in Michoacán, where they usually are found at elevations of less than 1000 meters in shaded places such as dense scrub forest, gallery forest, or tropical broad-leaf forest (Fig. 5).

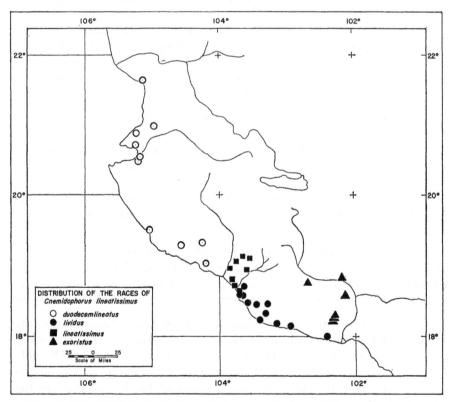

FIG. 5. Map showing distribution of the races of *Cnemidophorus lineatissimus* in México.

DIAGNOSIS.—In scutellation this species is intermediate between *guttatus* and *deppei; lineatissimus* has 108 to 164 dorsal granules at midbody, 28 to 48 femoral pores, and 5 to 11 preanal scales. The supraorbital semicircle series are complete in 78 per cent of the specimens in a sample from Nayarit; in other parts of the range relatively fewer individuals have the series complete, especially to the south, where along the coast of Michoacán the series never extend anteriorly beyond the posterior edge of the frontal. The largest male has a snout-vent length of 106 mm.

The dorsal color pattern of juveniles consists of nine or ten longitudinal stripes. In adults there are seven to nine stripes; the reduction is accomplished by the fusion of the vertebrals to form a single stripe, or the fusion of the paravertebrals with the vertebral to form a single stripe. A broad yellow middorsal stripe is characteristic of adult males of this species, except in some individuals from the northern part of the range in which the stripes are diffuse or absent. In adults, spots form in the lateral stripe and sometimes on the flanks and in the lower dorsolateral stripe; in some, these form vertical bars on the flanks. Adult males usually have pinkish blue throats, black gular collars, and blue bellies. In some the throat and collar are black.

GEOGRAPHIC VARIATION.—The characters of size, scutellation, and coloration vary independently throughout the range, resulting in different combinations of characters in various populations. The largest specimens are found in the south along the coast of Michoacán and in the Tepalcatepec Valley in Michoacán; the smallest adult males are from Nayarit (largest male from San Blas has a snout-vent length of 84 mm.).

The geographic variation in the number of dorsal granules at midbody presents a confusing picture (Fig. 6). The northern populations (coastal regions of Nayarit and Jalisco) have 132.9 ± 0.47 granules; the sample from the coast of Michoacán has 148.0 ± 1.24. Specimens from the lowlands of Colima (lying between Michoacán and Jalisco) have 120.0 ± 0.73 granules, and those from Tepalcatepec Valley have 122.4 ± 1.04. With respect to the number of femoral pores, those from Nayarit and Jalisco have 33.3 ± 0.25, and those from Colima have 33.3 ± 0.32. The southern populations have a greater number; those from the coast of Michoacán have 37.9 ± 0.38, and those from the Tepalcatepec Valley have 38.8 ± 0.48 (Fig. 7). The sample from Colima has an average of 6.7 preanal scales; that from the Tepalcatepec Valley, 6.5, whereas that from the coast of Michoacán has 7.4, and that from Jalisco and Nayarit has 7.5. The percentage of individuals having the supraorbital semicircle series complete varies greatly from one population to another; they are complete in 78 per cent of the specimens from Nayarit, 31 per cent from Jalisco, 11 per cent from Colima, 5 per cent from the Tepalcatepec Valley, and in none from the coast of Michoacán.

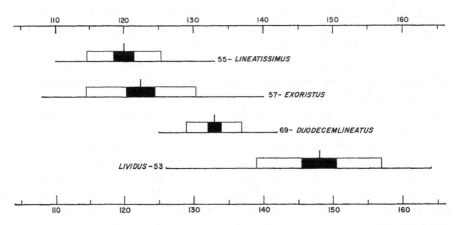

Fig. 6. Geographic variation in the number of dorsal granules in *Cnemidophorus lineatissimus*. See Figure 3 for explanation.

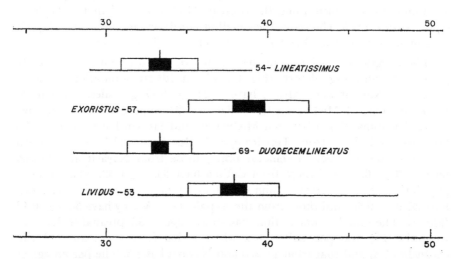

Fig. 7. Geographic variation in the number of femoral pores in *Cnemidophorus lineatissimus*. See Figure 3 for explanation.

In coloration the population on the coast of Michoacán and that in Nayarit and Jalisco are alike in that the stripes are diffuse or absent in large adults; furthermore, they are similar in having dark lateral fields and flanks, a character shared with the population in the Tepalcatepec Valley. The populations in Colima and the Tepalcatepec Valley are alike in having persistent stripes, but the former has nine stripes as adults, whereas the latter has only seven.

The combination of characters of scutellation and coloration in these samples leads us to the recognition of the four races of *lineatissimus* diagnosed below. Two of these—*lineatissimus* and *exoristus* (the Colima and Tepalcatepec Valley populations, respectively)—are much more alike in most characteristics than either is to the other populations, which, in turn, resemble one another more closely. However, *lineatissimus* ranges between the other populations. Although there is no evidence for intergradation between these forms, we have considered the four populations as subspecies of *lineatissimus*. Further investigation may show that two species are involved.

Cnemidophorus lineatissimus lineatissimus Cope

Cnemidophorus lineatissimus Cope, 1877, Proc. Amer. Philos. Soc., 17: 94.

Cnemidophorus deppei lineatissimus Cope, 1892, Trans. Amer. Philos. Soc., 17: 31.

Cnemidophorus deppii deppii, Burt, 1931, U. S. Natl. Mus. Bull., 154: 56–63 (part).

Cnemidophorus deppii lineatissimus, Hartweg and Oliver, 1937, Occ. Papers Mus. Zool. Univ. Michigan, 392: 2. Smith and Taylor, 1950, U. S. Natl. Mus. Bull., 199: 179.

SYNTYPES.—United States National Museum Nos. 32299–32314 from Colima, Colima, collected by John Xantus. Other specimens included in the original description by Cope (1877) are USNM 24937–40, part of the collection made in México by J. J. Major and shipped from Guadalajara. Smith and Taylor (1950) restricted the type locality to Colima, Colima.

DIAGNOSIS.—A moderate-sized race of *C. lineatissimus* characterized by a relatively low number of dorsal granules at midbody (average 120), a relatively low number of preanal scales (average 6.7), ten light longitudinal stripes in juveniles, vertical blue bars on the flanks, and fused vertebral stripes in adults.

DESCRIPTION.—The following description is based on 55 specimens from Colima. The largest male has a snout-vent length of 96 mm.; the largest female, 79 mm. Six males with snout-vent lengths of more than 90 mm. have tail lengths of 210 to 221 mm. and tail/body ratios of 2.25 to 2.43 (2.32). The smallest juvenile has a snout-vent length of 31 mm.

Scutellation: The number of dorsal granules at midbody varies from 110 to 133 (120.0 ± 0.73). Only two specimens have more than 129 granules. The number of femoral pores varies from 29 to 39 (33.3 ± 0.32). In the apex of the femora anterior to the enlarged preanals are one to five small scales; the total number of preanal scales varies from 5 to 9 (6.7). Of 53 specimens, 45 have the supraorbital semicircle series extending anteriorly only to the posterior edge of the frontal; in two others the series extend to the middle of the frontal, and in six the series are complete.

Coloration: Juveniles are black with a grayish olive head and ten pale cream longitudinal stripes; the vertebral stripes are noticeably narrower than the others. The upper surfaces of the limbs are dull brown with cream streaks and spots; the tail is pale blue. In adults the dorsal ground color is brown with a slight greenish tinge. The lateral dark field and the flanks are dark chocolate brown. Spots form in the pale blue lateral stripes, which are fragmented in some large adults, and expand laterally to form vertical bars on the flanks. The vertebral stripes are fused to form a pale yellow middorsal stripe. The other stripes are pale greenish yellow. The dorsal surfaces of the limbs are grayish brown, sometimes indistinctly marked with greenish cream spots. The upper surface of the tail is bluish gray. In adult males the chin is pinkish blue; the belly is bluish white, and there is a black gular collar. In old males the chin, throat, and chest are black, and the undersides of the limbs and tail are cream.

Specimens from near the city of Colima had in life an olive head, greenish brown dorsum with pale greenish yellow stripes, dark chocolate brown flanks with pale blue vertical bars, a dull bluish gray tail, and a pale bluish white belly in the males. Small adult males had a pale blue chin with a definite rose-pink cast.

ONTOGENETIC CHANGE IN COLOR PATTERN.—In this form there is no drastic metamorphosis of the striped juvenile color pattern. The dark fields become lighter with age; the lateral dark field and flanks change from black to dark chocolate brown, and the others change from black to a much lighter brown. The narrow area between the thin vertebral stripes begins to lighten anteriorly (Fig. 13). The fusion of the vertebral stripes progresses posteriorly, resulting in a rather broad middorsal stripe in place of the pair of thin vertebral stripes present in juveniles. Spots form in the lateral stripe; these expand to form vertical light bars on the flanks.

SEXUAL DIMORPHISM.—Males attain a larger size than females and have a more nearly complete color pattern metamorphosis. Adult males have a pinkish blue or black chin, black gular collar, and a blue belly, whereas the females have a creamy white venter.

GEOGRAPHIC VARIATION.—Four specimens from Manzanillo (UIMNH 36096–9) have 138 to 156 (148) dorsal granules, but otherwise resemble typical *lineatissimus*. We are unable to offer an explanation of this variation; Zweifel (1959) found a similar difference in samples of *Cnemidophorus communis* from Manzanillo and the city of Colima. Four specimens from Boca de Apiza, Michoacán, have 117 to 126 dorsal granules at midbody; in size and coloration they appear to be immature individuals of *C. l. lineatissimus*. Several adult males have light blue spots on the flanks; in some these are connected with the lateral stripe by vertical bars. In two large males, spots are present in the lower dorsolateral stripe. Data from the Michoacán and Manzanillo specimens are excluded from Tables II and III.

COMPARISON WITH OTHER FORMS.—The relatively low numbers of dorsal granules and preanal scales seem to indicate a close relationship of *lineatissimus* with *exoristus*, the race in the Tepalcatepec Valley. These races differ primarily in color pattern; adult *exoristus* have the paravertebral stripes fused with the middorsal stripe, resulting in only seven stripes as compared with nine in *lineatissimus*. From *lividus*, the race on the coast of Michoacán, *lineatissimus* differs in having fewer dorsal granules at midbody (120.0 ± 0.73, as compared with 148.0 ± 1.24) and in several aspects of coloration; *lineatissimus* has light brown paravertebral dark fields, dorsal stripes present on the neck, and complete (or nearly so) lateral and dorsolateral stripes; on the other hand, *lividus* has dark brown paravertebral dark fields, stripes indistinct or absent on the neck, and fragmented lateral and lower dorsolateral stripes. From *duodecemlineatus*, *lineatissimus* differs in having fewer dorsal granules (120.0 ± 0.73, as compared with 132.9 ± 0.47) and more pronounced dorsal stripes in adults.

ECOLOGICAL NOTES.—Specimens of this form have been collected in dense scrub forest and in gallery forest. No other member of the *deppei* group occurs with it. However, it is sympatric with *Cnemidophorus communis* and *Ameiva undulata*; the latter is an inhabitant of shaded areas, whereas the former lives in open scrub forest.

DISTRIBUTION.—This race is known only from the coastal lowlands of Colima and extreme western Michoacán and from the plateau of Colima at an elevation of about 500 meters. Locality records (72 specimens) follow.

México: *Colima*: No other data, UMMZ 48101–2; 3 km. E of Colima, UMMZ 114781 (5); 6 km. E of Colima, UMMZ 114779 (3); 8 km. E of Colima, UMMZ 114780 (3); Manzanillo, UIMNH 36096–9; Pascuales, UMMZ 80124 (5); Paso del Río, UIMNH 36087–95; Periquillo, UMMZ 80128 (11), 80129–30; Puebla Juarez, UMMZ 115575 (3); Queseria, UMMZ 80121 (2); Río Salada, UMMZ 80127 (3); 8 km. SW of Tecomán, UMMZ 80123 (4), 80125 (2); 3–5 km. NW of Villa Alvarez, UMMZ 80122 (5), 80126 (5). *Michoacán*: Boca de Apiza, UMMZ 118059 (4).

Cnemidophorus lineatissimus exoristus, new subspecies[3]
(Pl. I)

HOLOTYPE.—University of Michigan Museum of Zoology No. 119338 from Rancho Santa Ana (4 km. by road northeast of San Salvador, 600 meters), Michoacán, México, one of a series collected by William E. Duellman, Jerome B. Tulecke, and John Wellman, June 24, 1958. Original number, WED 12987.

PARATOPOTYPES.—UMMZ 119317–119337.

DIAGNOSIS.—A race of *C. lineatissimus* characterized by a relatively low number of dorsal granules at midbody (average about 122), relatively few preanal scales (average 6.5), nine longitudinal light stripes in juveniles, seven longitudinal light stripes in adults (paravertebrals fused with vertebral), and pale blue vertical bars on the flanks in adults.

DESCRIPTION OF HOLOTYPE.—An adult male with a snout-vent length of 90 mm., a tail length (complete) of 224 mm., and a tail/body ratio of 2.49. The scutellation is typical of *lineatissimus*—three supraoculars, enlarged mesoptychials, and granular postantebrachials. The supraorbital semicircle series extend anteriorly almost to the posterior edge of the frontal. At midbody there are 126 dorsal granules; there are seven preanal scales and 42 femoral pores.

The top of the head, rostral, nasals, and upper parts of the postnasals and loreals are olive-brown. The lower parts of the postnasals and loreals, the preoculars, upper and lower labials are bluish cream. The upper surfaces of the limbs are dull brown; on the hind limbs are pale blue dashes arranged in transverse rows. The dorsal surfaces of the tail and the feet are bluish gray. The chin is blue medially and pinkish blue peripherally. There is a black gular collar. The belly is pale blue, darkest anteriorly. The undersurfaces of the tail, limbs, and posterior part of the belly are cream. The ground color of the flanks, the lateral dark fields, and the upper dorsolateral dark fields is dark chocolate brown; that of the lower dorsolateral dark field is a lighter brown. At midbody there are 23 granules between the enlarged ventrals and the lower edge of the pale blue lateral stripe. Spots are present in the lateral stripe; most of these extend ventrally to form pale blue vertical bars on the flanks. Spots are present in the dorsolateral stripes, which are greenish yellow. Anteriorly the paravertebral stripes are fused with the vertebral stripe to form a broad cream middorsal

---

[3] From the Greek ʼεξόριστος, exiled; here used in allusion to the isolated position of this population in the Tepalcatepec Valley.

stripe. Posteriorly the paravertebral stripes are represented by rows of greenish yellow spots partially fused with the vertebral stripe.

In life, the flanks and the lateral and upper dorsolateral dark fields were rich chocolate brown; the lower dorsolateral fields were brown with a slight greenish tinge. The middorsal stripe was creamy yellow; the dorsolateral stripes were greenish yellow, and the lateral stripes and vertical bars on the flanks were pale blue. The throat was pinkish blue, and the belly was pale blue.

DESCRIPTION OF THE SUBSPECIES.—The following description is based on 57 specimens. The largest male has a snout-vent length of 98 mm.; the largest female, 82 mm. Ten males with snout-vent lengths of more than 90 mm. have tail lengths of 208 to 228 mm. and tail/body ratios of 2.17 to 2.49 (2.32). The smallest juvenile examined has a snout-vent length of 40 mm.

Scutellation: The dorsal granules vary in number at midbody from 108 to 140 (122.4 $\pm$ 1.04). Eight specimens have fewer than 115, and eight have more than 130. Sometimes two or three small scales lie in the apex of the femora anterior to the enlarged preanal scales, but often there is only one. The total number of preanal scales varies from 5 to 8 (6.5). The femoral pores vary in number from 32 to 47 (38.8 $\pm$ 0.48). The supraorbital semicircle series are complete in three specimens; in the others the series extend no farther anteriorly than the posterior edge of the frontal.

Coloration: Juveniles are black with olive gray heads and pale blue tails. They have nine creamy white longitudinal stripes, of which the laterals are the widest and the paravertebrals the narrowest. Pale cream spots and streaks are present on the dorsal surfaces of the limbs. Every adult has a broad cream middorsal stripe, which usually is bordered by a dark brown field. The dorsolateral stripes are distinctly greenish. The lateral stripes and vertical bars are light blue. Although spots form in all the stripes, only the paravertebrals posteriorly fragment into rows of spots; the others persist as stripes. The stripes are distinct from the neck to the region above the insertion of the hind limbs. The color of the chin of adult males varies from rosy pink to pinkish blue. The spots on the forelimbs disappear in large adults, but those on the hindlimbs persist. In some individuals the spots on the hindlimbs form transverse rows; in others they are irregular.

ONTOGENETIC CHANGE IN COLOR PATTERN.—The metamorphosis of color pattern in this form involves a reduction from nine to seven longitudinal stripes (Fig. 13). In juveniles the paravertebral stripes are narrowly separated from the single vertebral stripe. The paravertebral stripes widen and fuse with the vertebral. This fusion takes place first anteriorly. Usually, before the paravertebrals fuse posteriorly, they fragment into rows of spots, which later either completely or partially fuse with the vertebral stripe. In very large individuals the paravertebrals and vertebral become diffuse posteriorly. In this manner a rather broad middorsal stripe is formed from three separate narrow stripes. In some subadults the vertebral stripe is absent or indistinct; in individuals without a vertebral stripe the paravertebrals broaden and fuse together. In subadults spots appear in the lateral and dorsolateral stripes; however, these stripes never fragment into rows of spots. In some individuals a row of spots is present on the flanks midway between the ventrals and the lateral stripe. These and the spots in the lateral stripe expand to form vertical blue bars on the flanks. In other individuals the vertical bars appear to have been formed solely by the expansion of the spots in the lateral stripe. In old individuals some of the spots in the dorsolateral stripes expand to connect with other stripes. The flanks and lateral dark fields change from black to rich chocolate brown. In old individuals the lower, and sometimes the upper, dorsolateral fields become brown with a slightly greenish tint; however, in younger individuals the upper dorsolateral field is rich chocolate brown.

SEXUAL DIMORPHISM.—As in the other races of *lineatissimus,* the males attain a greater size than females and have a more nearly complete change of color pattern. Females have creamy white ventral surfaces as compared with the pinkish throats, black collars, and blue bellies of the males.

GEOGRAPHIC VARIATION.—Little geographic variation in scutellation or in coloration is evident in the present series. Twelve specimens from 13 to 21 kilometers south of Arteaga are from the Pacific slopes of the Sierra de Coalcomán and not distantly separated from nor geographically isolated from *C. lineatissimus lividus* on the Pacific Coast near Playa Azul. These specimens not only more closely resemble *exoristus* in the Tepalcatepec Valley in coloration, but also in the number of dorsal granules. Those from south of Arteaga have 109 to 140 (124.5 ± 2.75) dorsal granules at midbody, as compared with 108 to 135 (121.9 ± 1.10) for 45 specimens from the Tepalcatepec Valley and 129 to 155 (143.3 ± 2.97) for eight specimens from the eastern part of the coast of Michoacán.

COMPARISON WITH OTHER FORMS.—*Cnemidophorus lineatissimus exoristus* closely resembles *C. l. lineatissimus* in scutellation, for they have, respectively, 122.4 ± 1.04 and 120.0 ± 0.73 dorsal granules at midbody, 38.8 ± 0.48 and 33.3 ± 0.32 femoral pores, and 6.5 and 6.7 preanal scales. The only great difference in scutellation is in the number of femoral pores. In coloration, however, *exoristus* differs in having only seven stripes in adults, whereas *lineatissimus* has nine. Both *lividus* and *duodecemlineatus* have more dorsal granules than *exoristus*; *lividus* has 148.0 ± 1.24, and *duodecemlineatus* has 132.9 ± 0.47. Furthermore, *lividus* has nine stripes in adults, and *duodecemlineatus* has indistinct dorsal stripes. Even though there is a difference in the number of stripes, *exoristus* resembles *lividus* more than *lineatissimus* in certain aspects of coloration. There is a similarity in the dark color of the flanks, the lateral dark fields, and especially the dark border of the middorsal light stripe. However, the stripes in *exoristus* are persistent (even though spots form in the stripes), not fragmented into rows of spots as in *lividus*.

From *Cnemidophorus deppei infernalis,* a sympatric form, *exoristus* may be distinguished most readily by its different color pattern; *infernalis* never has vertical bars on the flanks, nor spots forming in the stripes. Furthermore, the adult males of *infernalis* have black throats and bellies, not pink throats and blue bellies as in *exoristus*. Immature specimens may be distinguished by the number of granules between the lower edge of the lateral stripe and the enlarged ventrals; there are 18 to 23 granules in *exoristus* and 12 to 16 in *infernalis*. There is a significant average difference in the number of dorsal granules at midbody; *infernalis* has 88 to 120 (99.3 ± 0.56) and *exoristus* has 108 to 140 (122.4 ± 1.04).

ECOLOGICAL NOTES.—All specimens of *exoristus* have been taken in more or less open gallery forest along streams in areas otherwise supporting scrub forest. In these gallery forests there is almost continuous shade, a leaf litter, and few or no grasses and herbs. In this habitat at Limoncito, *exoristus* was found associated with *Ameiva undulata*. At Capirio, *C. deppei infernalis* inhabits the open scrub forest adjacent to the gallery forest inhabited by *exoristus*, which at this locality was found with *C. calidipes* and *C. sacki (auctorum)*.

DISTRIBUTION.—This race inhabits the above-mentioned environments in the Tepalcatepec Valley and at least the easternmost Pacific slopes of the Sierra de Coalcomán, and perhaps the lower Balsas Valley in Michoacán from elevations of 170 to 600 meters in the valley and 750 to 900 meters on the Pacific slopes. Possibly it ranges westward through the Ahuijullo depression in southeastern Jalisco, where it may intergrade with *C. l. lineatissimus*. Locality records (64 specimens) follow.

México. *Michoacán*: 13 km. S of Arteaga, UMMZ 119347–50; 21 km. S of Arteaga, UMMZ 119339–46; 25 km. S of Arteaga, UKMNH 29685, 29687–9, 29691, 29700; Capirio, UMMZ 112659–60, 114758 (2), 114772 (4), 119302–4; Río Tepecuate near Limoncito, UMMZ 119305–16, 119520; Santa Ana, UMMZ 119317–38.

Cnemidophorus lineatissimus duodecemlineatus Lewis, new combination

Cnemidophorus deppii deppii, Burt, 1931, U. S. Natl. Mus. Bull., 154: 56–63 (part).

Cnemidophorus deppii lineatissimus, Smith and Taylor, 1950, U. S. Natl. Mus. Bull., 199: 179 (part).

Cnemidophorus guttatus duodecemlineatus Lewis, 1956, Nat. Hist. Misc., Chicago Acad. Sci., 156: 1–5.

Cnemidophorus deppei duodecemlineatus, Zweifel, 1959, Amer. Mus. Nat. Hist., Bull. 117 (2): 64.

HOLOTYPE.—Museum of Natural History, College of Puget Sound, No. 7547, collected by Murray L. Johnson and Thomas H. Lewis at Cerro de la Contaduria, San Blas, Nayarit, México.

DIAGNOSIS.—A small race of *lineatissimus* (less than 100 mm. snout-vent length) characterized by a moderate number of dorsal granules at midbody (average about 133), ten light stripes in juveniles, and indistinct or no stripes (except laterals) in adults.

DESCRIPTION.—The measurements given below are from the entire sample of 69 specimens, whereas the description of scutellation and coloration is based on 28 topotypes. The largest male has a snout-vent length of 92 mm.; the largest female, 72 mm. Four males with snout-vent lengths of more than 70 mm. have tail lengths of 162 to 198 mm. and tail/body ratios of 2.19 to 2.45 (2.32). The smallest juvenile examined has a snout-vent length of 33 mm.

Scutellation: The number of dorsal granules at midbody varies from 126 to 142 (133.9 ± 0.72); four specimens have less than 130, and only two have more than 140. There may be three or four small scales in the apex of the femora anterior to the enlarged preanal scales; the number of preanal scales varies from 6 to 8 (7.4). The femoral pores vary in number from 28 to 37 (32.3 ± 0.38). The supraorbital semicircle series are complete in 23 specimens; in four the series reach only to the posterior edge of the frontal, and in one they extend to the posterior edge of the second supraocular.

Coloration: Juveniles have a black dorsal ground color, a grayish olive head, a blue tail, and ten creamy white longitudinal stripes. There are irregular cream dashes and spots on the dark brown limbs. In females and immature males the lateral light stripe is broad and cream, and without spots. The other stripes are dull cream and are not distinct against the light brown dorsum. The lateral dark field is chocolate brown and noticeably darker than the dorsal dark field and somewhat darker than the flanks. Adult males have an olive brown head and greenish brown dorsum with faint greenish yellow stripes. In some the stripes (except the laterals) are absent, and in others there is a broad middorsal stripe at least anteriorly. Bluish white spots form in the lateral stripes; in some the lateral stripe is partially fragmented, and the spots are expanded to form vertical bars on the flanks.

In life the dorsum is greenish brown, and the stripes are greenish yellow. The lateral spots are pale blue. Males have a black throat and light blue belly.

ONTOGENETIC CHANGE IN COLOR PATTERN.—The metamorphosis of color pattern primarily consists of a fading of the dorsal stripes and a change from a black to a greenish brown ground color (Fig. 14). Juveniles have ten longitidinal light stripes. A middorsal light stripe appears between the vertebral stripes. Anteriorly in adults the vertebrals usually are fused with the middorsal stripe to form a broad middorsal light stripe, which usually is a lighter yellow than the other stripes. In old individuals the stripes become diffuse; in some even the broad middorsal stripe is absent. The lateral stripes either persist or partially break down into rows of spots. These sometimes are connected with spots on the flanks to form pale blue vertical bars.

SEXUAL DIMORPHISM.—Adult males have black throats and blue bellies spotted with black. Females have creamy white ventral surfaces. No females have spots formed in the lateral stripes nor on the flanks. Thus, in coloration they resemble immature males.

GEOGRAPHIC VARIATION.—Available specimens from Nayarit (32) have 126 to 142 (134.1 ± 0.66) dorsal granules at midbody as compared with 36 specimens from Jalisco, which have 125 to 141 (131.8 ± 0.61), and one from Colima, which has 139. The number of femoral pores varies from 28 to 37 (32.4 ± 0.35) in specimens from Nayarit and from 30 to 38 (34.2 ± 0.30) in those from Jalisco; the specimen from Colima has 34. The number of preanal scales varies from 6 to 9 (7.4) in specimens from Nayarit and from 6 to 10 (7.6) in those from Jalisco; the specimen from Colima has 6. Seventy-eight per cent of the specimens from Nayarit have the supraorbital semicircle series complete; they are complete in only 31 per cent of the specimens from Jalisco. There is little variation in coloration, except for seven of the 30 specimens from Puerto Vallarta, Jalisco (UIMNH 41394–41400). In most details of color pattern these specimens are like *C. l. lividus,* but the number of dorsal granules varies from 130 to 141 (134.6), well within the range of "normal" *duodecemlineatus.* Possibly *lividus* and *duodecemlineatus* are specifically distinct and occur sympatrically at least in the vicinity of Puerto Vallarta. On the other hand, the general nature of their distribution, together with their morphological and coloration differences (as well as similarities), suggests a subspecific relationship.

COMPARISON WITH OTHER FORMS.—The indistinct stripes in adults immediately distinguish this form from the other races of *lineatissimus*. Both *lividus* and *duodecemlineatus* have more preanal scales (average 7.4 in *lividus* and 7.5 in *duodecemlineatus*) than the other races (average 6.7 in *lineatissimus* and 6.5 in *exoristus*).Likewise, *duodecemlineatus* has more dorsal granules at midbody than *lineatissimus* or *exoristus* (132.9 ± 0.47 in *duodecemlineatus*, as compared with 120.0 ±0.73 in *lineatissimus* and 122.4 ± 1.04 in *exoristus*), but fewer than *lividus* (148.0 ± 1.24).

ECOLOGICAL NOTES.—This small form has been collected in oil palm groves and tropical semi-deciduous broad-leaf forest, where it appears to prefer shaded areas to open sunny ones. Occurring sympatrically with it in Nayarit is *C. sacki huico,* and in Jalisco and Colima are *C. communis* and *Ameiva undulata*.

DISTRIBUTION.—This race occurs on the Pacific Coast and foothills to elevations of about 600 meters from San Blas, Nayarit, southeastward to northwestern Colima. No intergrades between *duodecemlineatus* and *lineatissimus* are known. Locality records (81 specimens) follow.

México: *Colima*: Ejido de Tepextle, 6 km. ENE of Manzanillo, UMMZ 115585. *Jalisco*: Bahía Chamela, AMNH 62779–80, UMMZ 84246; 3 km. N of La Resolana, UMMZ 102053–4; Puerto Vallarta, AMNH 15748–50, 15753–72, UIMNH 41394–400; 8 km. S of Purificación, UKMNH 27198; Río Real, Penas, Bahía Banderas, AMNH 62781; 8 km. SW of Tecomate, UKMNH 29696–7. *Nayarit*: Bahía Banderas, UMMZ 84247 (2); 10 km. SE of Las Varas, UKMNH 29690, 29692; San Blas, AMNH 15843–7, 15849–50, 15873–4, UKMNH 27729, 27741, 29693–5, 29698–9, UMMZ 104737 (2), 112654 (6), 114766 (10); San José de la Conde, UMMZ 102055; Sayulita, UMMZ 113066.

Cnemidophorus lineatissimus lividus, new subspecies[4]
(Pl. I)

Cnemidophorus guttatus immutabilis, Peters, 1954, Occ. Papers Mus. Zool. Univ. Mich., 554: 18.

HOLOTYPE.—University of Michigan Museum of Zoology No. 119472, from Maruata (18° 17′ N, 103° 20′ W, sea level), Michoacán, México, collected by William E. Duellman on July 15, 1951. Original number, WED 5124.

PARATOPOTYPES.—UMMZ 104735, 104740, 105136, 119473–119475.

DIAGNOSIS.—A race of *lineatissimus* characterized by a high number of

---

[4] Latin, *lividus,* bluish or to become blue; here alluding to the bluish color of the feet and flanks of adults of this form.

dorsal granules at midbody (average 148), supraorbital semicircle series usually not extending anteriorly beyond the posterior edge of the frontal, and a broad middorsal stripe bordered by dark brown and vertical blue bars on the flanks in adults.

DESCRIPTION OF HOLOTYPE.—An adult male with a snout-vent length of 91 mm., a tail length (complete) of 208 mm., and a tail/body ratio of 2.28. The scutellation is typical of *lineatissimus*—three supraoculars, enlarged mesoptychials, and granular postantebrachials. The supraorbital semicircle series extend anteriorly nearly to the posterior edge of the frontal. At midbody there are 155 dorsal granules; there are ten preanal scales and 38 femoral pores.

The top of the head is light olive brown; this color extends onto the rostral, nasals, postnasals, anterior upper labials, and the mental. The loreals, preoculars, and other labials are bluish gray. The upper surfaces of the hind limbs are brownish gray; the hind limbs are spotted with bluish white. The proximal one-fourth of the tail is brownish gray; posteriorly it is bluish gray. The upper surfaces of the feet are dark bluish gray. The chin is pinkish blue, and the belly is bluish white. The lower surfaces of the limbs and tail are bluish cream. There is a broad black gular collar. Extending posteriorly from a point immediately behind the interparietal to a point above the insertion of the hind limbs is a broad cream stripe (6 granules in width at midbody) bordered on either side by a narrow chocolate brown dark field. Lateral to this is a row of faint spots representing the paravertebral stripe. With the exception of the middorsal stripe, anteriorly the stripes suffuse with the grayish tan ground color. Separating the paravertebral and the upper dorsolateral stripes is a grayish brown dark field. The upper dorsolateral stripe is complete and shows no indication of forming spots, whereas spots are present in the lower dorsolateral stripe. The lower dorsolateral and lateral dark fields are dark chocolate brown. There are 27 granules between the enlarged ventrals and the lower edge of the bluish white lateral stripe, which is partially fragmented into large round spots. The flank below the lateral stripe is brownish black. Along the lower edge of the flank is a row of bluish white spots; midway between the enlarged ventrals and the lateral stripe is a similar row of spots.

In life the top of the head was light olive tan; the sides of the head and neck were bluish gray. The dorsum anteriorly was greenish gray, and the dark fields and flanks were deep chocolate brown. The middorsal stripe was bright yellow; the paravertebrals and upper dorsolateral stripes were pale cream with a greenish tinge, and the lower dorsolaterals, laterals, and spots on the flanks were light blue. The chin was pink; the gular region was black, and the belly was pale blue.

DESCRIPTION OF THE SUBSPECIES.—The following description is based on 54 specimens from the coastal lowlands and foothills in Michoacán. The largest male has a snout-vent length of 106 mm.; the largest female, 89 mm. Nine males with snout-vent lengths of more than 90 mm. have tail lengths of 205 to 250 mm. and tail/body ratios of 2.13 to 2.36 (2.24). The smallest juvenile examined has a snout-vent length of 44 mm.

Scutellation: The dorsal granules at midbody vary in number from 126 to 164 (148.0 ± 1.24). Only two specimens have fewer than 135, and only six have more than 160. In some individuals the enlarged preanal scales are preceded by only two or three scales in the apex of the femora; in others there are as many as seven. The total number of preanal scales varies from 5 to 11 (7.4). The femoral pores vary in number from 32 to 48 (37.9). The supraorbital semicircle series do not extend anteriorly beyond the posterior edge of the frontal in any specimen.

Coloration: Juveniles have a black body fading to gray along the lower edges of the flanks. With the exception of the lateral ones, which are bluish white, the longitudinal stripes are pale yellow. The tail is light blue. Pale cream dashes or streaks are present on the upper surfaces of the limbs. In some immature individuals there is a single vertebral stripe, but in most there is a pair of vertebral stripes. Individuals collected in the dry season have a light brown dorsum with greenish yellow stripes. All adults have the broad middorsal yellow stripe and at least a partially fragmented lateral stripe. Anteriorly, and to a lesser extent posteriorly, the dorsolateral and paravertebral stripes are diffuse. Adult males have the spots on the flanks connected with the spots formed by the fragmentation of the lateral stripe; in large males these form vertical bars on the flanks. Small subadults often have the light stripes narrowly outlined by black.

ONTOGENETIC CHANGE IN COLOR PATTERN.—The metamorphosis of color pattern in this form involves a replacement of the lateral stripe by spots, the suffusion and loss of the stripes anteriorly, and the development of a broad middorsal light stripe by the fusion of the vertebral stripes, thus resulting in an adult lizard with nine stripes and rows of spots (Fig. 14). Juveniles have ten distinct longitudinal stripes, of which the lateral ones are the broadest and the vertebral pair the narrowest and fused anteriorly and posteriorly. With increased size and age the dark field between the vertebral stripes lightens, and the vertebral stripes become broader, until the entire area between the vertebral stripes is yellow, forming in the place of the orginal pair of light stripes and dark field a single broad yellow stripe. Apparently the formation of the solid broad middorsal stripe begins just back of the head and progresses posteriorly. In small adults spots appear on the flanks; these are in two rows—one just above the ventrals, and the other midway between the ventrals and lateral stripe. Later, spots form in the lateral stripe. These spots expand and fuse with those on the flanks to form vertical bars. In old males spots form in the lower dorsolateral stripe; the lateral stripe fragments, and the vertical bars expand to reach the lower dorsolateral stripe. In subadults the stripes are faint anteriorly, and in very large specimens the paravertebrals and upper dorsolaterals may be faint or absent. The light streaks on the forelimbs disappear in adults; those on the hind limbs break into spots.

SEXUAL DIMORPHISM.—Males, which attain a greater size than females, have a more advanced metamorphosis of color pattern. Females seldom lose the stripes anteriorly and often do not have spots formed in the lower dorsolateral stripes, nor do they have the colored throats and bellies of the adult males.

GEOGRAPHIC VARIATION.—The range of this form includes the coastal lowlands and foothills of Michoacán, an airline distance of about 200 kilometers. If the 26 specimens from the western part of this range (Boca de Apiza to La Placita) are compared with the 19 specimens from the middle of the range (vicinity of Maruata and Pómaro) and the eight from the eastern part (Río Nexpa to Playa Azul), the following variation is observed in the number of dorsal granules at midbody: those from the west have 137 to 164 (147.6 ± 1.51); those from the middle of the range have 126 to 164 (148.4 ± 2.06); those from the east have 129 to 155 (143.3 ± 2.97). The number of femoral pores in the same series varies as follows: those from the west have 33 to 43 (36.1 ± 0.57); those from the middle of the range have 34 to 46 (38.8 ± 0.62); those from the east have 38 to 48 (41.9 ± 0.77). The same series analyzed for variation in the number of preanal scales show the following: those from the west have 5 to 10 (7.2); those from the middle of the range have 6 to 11 (8.2); those from the east have 5 to 8 (6.8). No geographic variation in color pattern is apparent.

COMPARISON WITH OTHER FORMS.—In certain aspects of coloration *C. l. lividus* resembles both *exoristus* and *duodecemlineatus*; it differs from *exoristus* in having nine instead of seven stripes in adults and in having the lateral and sometimes the lower dorsolateral stripes fragmented. From *duodecemlineatus* it differs in having a bolder color pattern and persistent stripes (at least the vertebral and laterals), and from both of these forms *lividus* differs in having more dorsal granules at midbody (148.0 ± 1.24 in *lividus*, as compared with 122.4 ± 1.04 in *exoristus* and 132.9 ± 0.47 in *duodecemlineatus*). From *lineatissimus*, *lividus* differs in having the dark paravertebral and lateral fields and in having more dorsal granules (*lineatissimus* has 120.0 ± 0.73). From *C. d. deppei*, a sympatric species, *lividus* differs in having more dorsal granules (*deppei* has fewer than 120) and vertical bars on the flanks; *deppei* sometimes has the lateral stripe fragmented into a row of blue spots, but these are not expanded into vertical bars. Juveniles may be distinguished by the numbers of granules between the enlarged ventrals and the lower edge of the lateral stripe; in *lividus* there are more than 20, and in *deppei* 17 or less.

Superficially *lividus* resembles *C. guttatus flavilineatus* in Chiapas; the latter has a lighter flank, only seven longitudinal stripes, and 158.8 ± 1.53 dorsal granules at midbody.

ECOLOGICAL NOTES.—These lizards have been collected in dense scrub forest and in semi-deciduous broad-leaf tropical forest. They have been found in the scrub growth above the strand on beaches. This form usually is found in or close to shade, where it sometimes is seen with *Ameiva undulata,* a species characteristic of the more dense forest.

DISTRIBUTION.—This form has been collected only on the Pacific Coast and foothills of the Sierra de Coalcomán in Michoacán, to elevations of about 300 meters. It is doubtful if its range extends southeastward across the Río Balsas into Guerrero, where it apparently is replaced by *Cnemidophorus guttatus immutabilis*. Locality records (54 specimens) follow.

México: *Michoacán*: Barranca de Bejuco, UMMZ 104736 (2), 118058 (2); Boca de Apiza, UMMZ 104525, 105127; Coahuayana, UMMZ 114756 (3); 9 km. S of Coahuayana, UMMZ 104526 (2), 105128; El Ticuiz, UMMZ 114754 (4); 4 km. E of El Ticuiz, UMMZ 114755 (3); La Placita, UMMZ 104527–9, 105129, 114757 (7); Maruata, UMMZ 104735, 104740, 105136, 119472–5; Motín del Oro, UMMZ 105134; Ostula, UMMZ 105130 (2), 105131, 105132 (2); Playa Azul, UMMZ 112665 (4); Playa Cuilala, UMMZ 104741 (2); Pómaro, UMMZ 105137 (2); Salitre de Estopila, UMMZ 105133, 105135; San Pedro Naranjestila, UMMZ 114387.

Cnemidophorus guttatus Wiegmann

Cnemidophorus guttatus Wiegmann, 1934, Herpetologia Mexicana, pp. 27, 29.

DISTRIBUTION.—This species ranges from northern Veracruz in eastern México and from the Río Balsas in western México southward to the Isthmus of Tehuantepec, and thence eastward along the Pacific lowlands and foothills of Chiapas and into the upper Cintalapa Valley in Chiapas (Fig. 8). It inhabits shaded areas at elevations usually less than 1000 meters, but does not live in dense forest.

DIAGNOSIS.—This is the largest species in the group; adult males attain snout-vent lengths in excess of 140 mm. This species has 142 to 208 dorsal

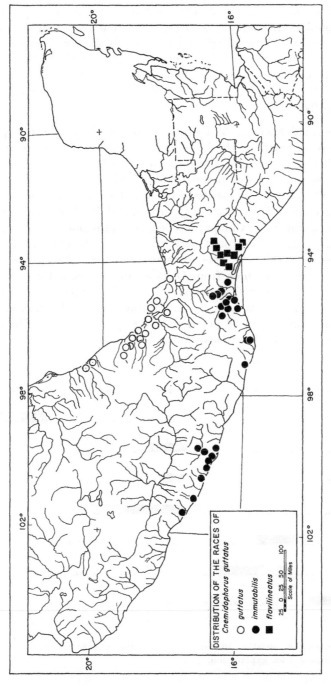

Fig. 8. Map showing distribution of the races of *Cnemidophorus guttatus* in México.

granules at midbody, 32 to 52 femoral pores, and 5 to 11 preanals. In different populations from 19 to 87 per cent of the individuals may have complete supraorbital semicircle series. The basic dorsal color pattern consists of seven or eight light stripes (upper dorsolaterals absent) or rows of spots on a brown dorsum; the lateral field is darker than the rest of the dorsum. Adult males have a black gular collar, a buff or orange throat, and a blue belly.

GEOGRAPHIC VARIATION.—With respect to size the southern population in Chiapas is the smallest; the largest male from Chiapas has a snout-vent length of 113 mm. Likewise, the population in Chiapas has fewer dorsal granules at midbody, fewer preanal scales, and fewer femoral pores than the others. Those from Chiapas have 158.8 ± 1.53 granules; there are 179.4 ± 2.29 in Oaxaca, 175.8 ± 1.22 in Guerrero, and 199.3 ± 1.11 in Veracruz (Fig 9). The sample from Chiapas has 38.2 ± 0.42 femoral pores; that from Oaxaca, 45.2 ± 0.60; that from Guerrero, 41.8 ± 0.40; and that from Veracruz, 44.2 ± 0.75 (Fig 10). Specimens from Veracruz have an average of 8.8 preanal scales, and those from Oaxaca and Guerrero have an average of 8.7, whereas those from Chiapas have 7.5. In the sample from Veracruz 87 per cent of the specimens have the supraorbital semicircle series complete; the series are complete in 29 per cent of the specimens from Chiapas and in 19 per cent of those from Oaxaca and Guerrero.

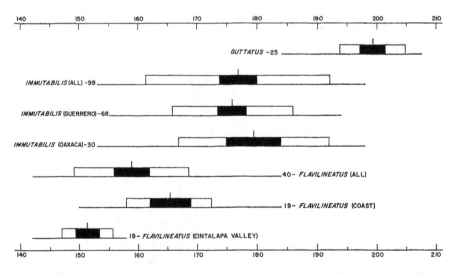

FIG. 9. Geographic variation in the number of dorsal granules in *Cnemidophorus guttatus*. See Figure 3 for explanation.

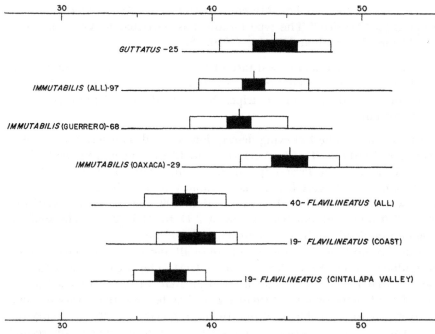

FIG. 10. Geographic variation in the number of femoral pores in *Cnemidophorus guttatus*. See Figure 3 for explanation.

Juveniles from Veracruz are brown with a lateral light stripe; juveniles from the rest of the range are black with seven or eight longitudinal light stripes. In the specimens from Veracruz rows of spots develop on the dorsum; in those from Oaxaca and Guerrero the longitudinal stripes of the juveniles fragment into rows of spots. In the specimens from Chiapas the vertebral stripes fuse into a broad midorsal light stripe, and all of the stripes are persistent.

On the basis of the differences in coloration which are fairly closely correlated with differences in scutellation three races are recognized. These are diagnosed below.

Cnemidophorus guttatus guttatus Wiegmann

Cnemidophorus guttatus Wiegmann, 1834, Herpetologia Mexicana, pp. 27, 29. Burt, 1931, U. S. Natl. Mus. Bull., 154: 66–74.

Cnemidophorus unicolor Cope, 1877, Proc. Amer. Philos. Soc., 17: 93 (USNM 30385; type locality "West Tehuantepec," Oaxaca, México; collected by Francis Sumichrast).

Cnemidophorus guttatus guttatus, Cope, 1892, Trans. Amer. Philos. Soc., 17: 32. Gadow, 1906, Proc. Zool. Soc. London, pp. 309, 325–326. Hartweg and Oliver, 1937, Occ. Papers Mus. Zool. Univ. Michigan, 359: 3, Smith and Taylor, 1950, Bull. U. S. Natl. Mus., 199: 179–180,

HOLOTYPE.—Zoologisches Museum Berlin, No. 887, collected by F. Deppe from "Mexico." The type locality was restricted to Veracruz, Veracruz, México, by Smith and Taylor (1950: 180).

DIAGNOSIS.—A moderate-sized race of *C. guttatus* with a high number of dorsal granules at midbody (average about 199); young with only lateral stripes present; adults without stripes, but instead rows of light spots on a brown dorsum.

DESCRIPTION.—The following description is based on a series of 13 specimens from Lerdo de Tejada, Veracruz, and 12 specimens from west of the city of Veracruz. The largest specimen is a male with a snout-vent length of 129 mm.; the largest female has a snout-vent length of 105 mm. Five males with snout-vent lengths of more than 100 mm. have tail lengths of 228 to 270 mm. and tail/body ratios of 2.11 to 2.23 (2.17). The smallest juvenile examined has a snout-vent length of 49 mm.

Scutellation: The dorsal granules are small and vary in number around the midbody from 184 to 208 (199.3 ± 1.11). The preanal scales vary in number from 7 to 10 (8.8); the enlarged preanal scales are preceded in the apex of the femora by several small scales. The femoral pores vary in number from 38 to 48 (44.2 ± 0.75). In most specimens (92 per cent) the supraorbital semicircle series is complete to the anterior edge of the second supraocular, which is separated from the first by a row of granules continuous with the row of granules separating the supraoculars from the superciliaries. Thus, in these specimens the second and third supraoculars are completely surrounded by granules. In the other specimens the supraorbital semicircle series extends anteriorly only to the posterior edge of the frontal or to about the middle of the frontal.

Coloration: In small individuals the entire dorsal surface of the body, head, and limbs is light brown; a narrow lateral light stripe extends from the upper edge of the ear to the anterior edge of the hind limb. In some specimens this stripe is broken into a series of dashes. Large individuals have a brown dorsum with somewhat lighter flanks. The lateral stripe or row of spots is indistinct. There are 30 to 34 granules between the enlarged ventrals and the lower edge of the lateral stripe. The lower dorsolateral stripe is present as a faint interrupted line or as a row of spots; the paravertebral and vertebral stripes are represented by rows of rather small light spots. Sometimes a faint row of spots is present between the vertebral rows. The lateral dark field always is slightly darker brown than the rest of the dorsum. The top of the head and the upper surfaces of the limbs and tail are olive brown and unmarked. The sides of the head are bluish gray. The throats of adult males are tan; there is a narrow bluish black gular collar.

The belly is bluish white medially and dark blue or black with round bluish white spots laterally and on the anterior and ventral surfaces of the hind limbs. The underside of the tail is bluish cream.

In life the dorsum is sandy brown sometimes tinged with olive; the lateral dark field is a dark chocolate brown, and the flanks are light olive-gray or olive-tan. The throats of adult males are buff and the bellies light blue medially and bluish black laterally. In juveniles the sides of the proximal half and the entire posterior half of the tail are pale blue.

ONTOGENETIC CHANGE IN COLOR PATTERN.—Insofar as known, the juveniles of *C. g. guttatus* do not have a black dorsal ground color, as is characteristic of the young of other forms in this group. Also, with the exception of the lateral one, the juveniles have no longitudinal light stripes. With increased size and age light spots appear on the dorsum (Fig. 15). These are small and appear first in the lines normally occupied by the lower dorsolateral stripes. Later, spots appear in paravertebral rows and sometimes in very large individuals in a vertebral row. The spots in old individuals are larger and less distinct than in smaller ones. With increased size the lateral stripe fragments into a row of spots, except posteriorly, where a stripe usually persists. Thus, the metamorphosis of color pattern in this form consists of the development of light spots; these are formed (with the exception of the lateral rows) without the fragmentation of previously existing longitudinal stripes.

SEXUAL DIMORPHISM.—Aside from the somewhat larger size attained by the males, the only noticeable sexual dimorphism is found in coloration. Correlated apparently with the larger size of the males is the more nearly complete metamorphosis of color pattern; the degree of dorsal spotting reached by the largest females is never as complete as that in the largest males. The buff throat, black collar, and blue belly distinguish adult males from females, which have a completely white or bluish white venter.

GEOGRAPHIC VARIATION.—Five specimens from Rodriguez Clara (UIMNH 36069–73) resemble *C. g. flavilineatus* in color pattern and in the number of dorsal granules at midbody (156–170); seven from San Gerónimo (UIMNH 36062–8) have stripes anteriorly like *immutabilis*. Such variants cannot be explained with certainty; perhaps after the differentiation of the three now recognized races of *guttatus* there was dispersal and subsequent isolation of small populations to form a mosaic in the isthmian region. Data from these variants are excluded from Tables II and III.

COMPARISON WITH OTHER FORMS.—From the other races of *C. guttatus*, this form may be distinguished principally by the different metamorphosis of color pattern. The juveniles of the Pacific Coast races of *guttatus* are black with at least seven light longitudinal stripes, which with age partially fragment into rows of spots or remain as stripes throughout life, whereas the young of *guttatus* are brown with only a lateral light stripe; with age the spots appear in rows on the dorsum. Both of the Pacific Coast races have fewer dorsal granules than *guttatus* (199.3 $\pm$ 1.11, as compared with 176.8 $\pm$ 1.54 for *immutabilis* and 158.8 $\pm$ 1.53 for *flavilineatus*). The supraorbital semicircle series is complete in only 30 per cent of specimens of *flavilineatus* and in 17 per cent of specimens of *immutabilis* as compared with 87 per cent of *guttatus*.

From the Gulf Coast *deppei*, *guttatus* may be distinguished by its greater maximum size (129 mm. as compared with 82 mm.), greater number of granules around the body (more than 180 as compared with fewer than 125), and in having juveniles without stripes and adults with spotted backs, contrasted with persistent stripes in *deppei*.

ECOLOGICAL NOTES.—*Cnemidophorus guttatus guttatus* inhabits brushy situations, including clumps of bushes in savannas, stream banks in open scrub forest, and the dense brush on the leeward sides of coastal dunes. In forests it is replaced by *Ameiva undulata* and in the open scrub forest and savanna grasslands by *Cnemidophorus deppei deppei*.

DISTRIBUTION.—This form is definitely known only from the state of Veracruz, where it is found in the habitats described above in the lowlands and foothills, usually at elevations of less than 1000 meters, from Tecolutla to Coatzacoalcos. Although there has been considerable collecting to the north of Tecolutla, principally near Tuxpan and Tampico, as yet this lizard has not been reported from there; it probably also occurs along the coastal areas of extreme eastern Veracruz and Tabasco. One specimen (UMMZ 88653), purportedly from San Diego, near Tehuacan, Puebla, doubtlessly bears incorrect data, for this locality is at an elevation of about 1800 meters in the upper Balsas Basin. Locality records (175 specimens) follow.

México: *Veracruz*: Alvarado, UMMZ 57002; 6 km. ESE of Alvarado, UMMZ 118212; 26 km. ESE of Alvarado, UIMNH 39300–3; Boca del Río, UIMNH 33883; 5 km. SW of Boca del Río, UKMNH 24434; 10 km. S of Boca del Río, UMMZ 118724 (2); Cempoala, UIMNH 26034–5; Coatzacoalcos, USNM 47525–6; near Jalapa, UMMZ 69267; 22 km. E of Jalapa, UMMZ 95100; 5 km. NW of Lerdo de Tejada, UMMZ 114756 (13); Matacabresto, UIMNH 36522–3; Palma Sola, UIMNH 36521; Plan del Río, UMMZ 102051, 102052 (4), 105622 (6); Piedras Negras, UMMZ 84515 (2), 88651 (12), 88652 (30); Puente Nacional, UIMNH 21843–5, 36074, UKMNH 24431–2, 24435, UMMZ 69409, 105616 (4); Punta Colorado, UMMZ 89323 (2); Río Blanco, 20 km. WNW of Piedras Negras, UKMNH 23312–3; Rodriguez Clara, UIMNH 36069–73; San Gerónimo, UIMNH 36062–8; Tecolutla, UIMNH 4169; 3 km. S of Tecolutla, UIMNH 3950; 16 km. S of Tecolutla, UIMNH 3851–68; Tierra Colorado, UIMNH 36058–61; 4 km. W of Tierra Colorado, UMMZ

95105 (9); 18 km. W of Veracruz, UMMZ 95102–3; 22 km. W of Veracruz, UMMZ 95106 (6); 35 km. W of Veracruz, UMMZ 99915 (7); 80 km. W of Veracruz, UMMZ 99916 (2); 8 km. SW of Veracruz, UMMZ 95104 (11); 6 km. S of Veracruz, UMMZ 95101.

Cnemidophorus guttatus immutabilis Cope

Cnemidophorus immutabilis Cope, 1877, Proc. Amer. Philos. Soc., 17: 93.

Cnemidophorus microlepidopus Cope, 1877, *Ibid.* (USNM 30187; type locality "West Tehuantepec," Oaxaca, México; collected by Francis Sumichrast. Type locality restricted to city of Tehuantepec [Smith and Taylor, 1950: 180]).

Cnemiodphorus guttatus immutabilis Cope, 1892, Trans. Amer. Philos. Soc., 17: 31. Gadow, 1906, Proc. Zool. Soc. London, pp. 309, 326–327. Hartweg and Oliver, 1937, Occ. Papers Mus. Zool. Univ. Michigan, 359: 3–7. Smith and Taylor, 1950, Bull. U. S. Natl. Mus., 199: 180.

Cnemidophorus guttatus striata Gadow, 1903, Proc. Royal Soc. London, 72; 115 (No type designated; type locality "Isthmus of Tehuantepec," restricted to Tehuantepec, Oaxaca, México, by Smith and Taylor, 1950: 180).

Cnemidophorus striatus Gadow, 1905, Proc. Zool. Soc. London, p. 195.

Cnemidophorus guttatus, Burt, 1931, Bull. U. S. Natl. Mus., 154: 66–74 (part).

HOLOTYPE.—United States National Museum No. 30141, collected by Francis Sumichrast from "West Tehuantepec," Oaxaca. The type locality was restricted to the city of Tehuantepec, Oaxaca, by Smith and Taylor (1950: 180).

DIAGNOSIS.—A large race of *C. guttatus* with a moderate number of dorsal granules at midbody (average about 177), young black and usually with seven longitudinal light stripes; adults with spots partly replacing the stripes.

DESCRIPTION.—The following description is based on a series of 30 specimens from the vicinity of Tehuantepec, Oaxaca. The largest male has a snout-vent length of 145 mm.; the largest female, 115 mm. Eight males with snout-vent lengths of more than 100 mm. have tail lengths of 228 to 260 mm. and tail/body ratios of 2.11 to 2.28 (2.21). The smallest juvenile examined has a snout-vent length of 48 mm.

Scutellation: The dorsal granules vary in number around the midbody from 153 to 198 (179.4 ± 2.29). The enlarged preanal scales are preceded in the apex of the femora by several small scales; the total number of preanal scales from the apex to the posterior enlarged scale varies from 6 to 11 (8.7). The femoral pores vary in number from 38 to 52 (45.2 ± 0.60). The supraorbital semicircle series is complete in five specimens; in two others the series extends anteriorly to the middle of the frontal, and in 23 the series reach only to the posterior edge of the frontal.

Coloration: Small juveniles are black with six well-defined longitudinal yellow or cream stripes; below the lateral stripe the flanks are dark bluish gray. An irregular middorsal or vertebral stripe usually is present; this sometimes is split, leaving a narrow black line between a pair of brownish vertebral stripes. There is only one pair of dorsolateral stripes. The top of the head is dark olive brown, and the sides of the head are brownish cream. The tail is bluish gray above and cream below. The upper surfaces of the thighs are streaked with cream. In large individuals the stripes are replaced by spots posteriorly. The dorsal ground color is dull brown or greenish brown; the lateral dark field is dark brown, and the flanks below the lateral stripe are gray. The head, except for the labials which are gray, is olive brown. The hind limbs and proximal part of the tail are flecked with yellow or greenish yellow. The ventral coloration of adult males consists of gray labials, a pinkish tan throat, a black gular collar, and a dull bluish gray belly and hind limbs. Laterally, on the belly, are bluish white spots. The ventral surface of the tail is dull cream.

In life the adults have a dull brown dorsum often tinged with green; the lateral dark field is dark chocolate brown. The lateral stripe is bluish gray, and the flanks below this stripe are a darker gray. The other stripes and spots are pale yellow or greenish yellow. The throats of adult males are dull orange or orange-buff.

ONTOGENETIC CHANGE IN COLOR PATTERN.—The change in color pattern in this race involves the metamorphosis of a black lizard with six light stripes and a vertebral light area to a brown lizard with eight stripes, which subsequently are partly replaced by spots (Fig. 15). Large juveniles or small subadults show a splitting of the vertebral light area into a pair of vertebral stripes. Later, spots develop in all of the stripes. The lateral stripe assumes a bluish color and connects with several bluish vertical bars on the flanks. In large individuals the vertebral stripes normally fade completely, leaving only rows of spots in their place. The paravertebral and dorsolateral stripes persist anteriorly, but posteriorly they are represented only by rows of spots. The lateral stripe persists but fades into the bluish gray flank.

SEXUAL DIMORPHISM.—The development of the adult color pattern can be correlated closely with the size of the individual. Males, which attain a greater size than females, show a greater degree of development of the spotted dorsum; nevertheless, large females do have spots replacing the stripes, particularly posteriorly. The white or bluish white bellies of the females are in contrast with the dull bluish gray bellies of males. Females often have a faint dark gular collar and a pinkish cast to the throat.

GEOGRAPHIC VARIATION.—The series from Tehuantepec may be compared with a series of 68 specimens from Tierra Colorado and the vicinity
of Acapulco, Guerrero, in the northwestern part of the range. Although
the sample from Guerrero is larger, less variation is noted in the number
of granules around the midbody; these vary from 155 to 194 (175.8 ± 1.22)
in Guerrero as compared with 153 to 198 (179.4 ± 2.29) at Tehuantepec.
Specimens from Guerrero have slightly fewer femoral pores; there are 34
to 48 (41.8 ± 0.40) as compared with 38 to 52 (45.2 ± 0.60) in specimens
from Tehuantepec. Essentially the same proportion of specimens from both
regions have the supraorbital semicircles complete (17 per cent in Oaxaca
and 20 per cent in Guerrero).

Several individuals from the region of Tehuantepec have a color pattern
approaching that of *flavilineatus*; likewise, four from Guerrero resemble
flavilineatus in coloration. Of the Oaxacan specimens, one from Chivela
(UMMZ 114764) has the pale flanks and dark dorsum of *flavilineatus*, but
the vertebral stripe is split and spotted; this specimen has 180 granules
around the body and 46 femoral pores, both close to the averages for these
characters in the sample of *immutabilis* from Tehuantepec. Nine specimens (UIMNH 36111–9) from Totolapam, Oaxaca, have dark fields and
six or seven persistent stripes; otherwise they are typical *immutabilis*.
Fourteen specimens from Cacalutla, Guerrero (UMMZ 119133), have ten
narrow stripes. Individuals from extreme southeastern Oaxaca (vicinity
of Tapanatepec) appear to be much closer to *flavilineatus*; these are discussed under that form.

COMPARISON WITH OTHER FORMS.—*Cnemidophorus guttatus immutabilis* may be distinguished from the Gulf Coast race *guttatus* by the fewer
number of dorsal granules (176.8 ± 1.54 as compared with 199.3 ± 1.11 in
guttatus), by the proportionately few specimens with complete supraorbital
semicircle series (87 per cent of the specimens of *guttatus* have the series
complete), and by the persistence of stripes anteriorly in the adults, whereas
in *guttatus* the entire dorsum is spotted. From *flavilineatus*, *immutabilis*
may be distinguished by its greater number of dorsal granules (176.8 ± 1.54
as compared with 158.8 ± 1.53 in *flavilineatus*); furthermore, *flavilineatus*
has broad persistent stripes contrasting with the dark dorsal ground color.

From *Cnemidophorus deppei deppei*, a sympatric species, *immutabilis*
may be distinguished by its much larger maximum size (145 mm. as compared with 80 mm. in *deppei*), greater number of granules around the body
(more than 150 as compared with less than 115 in *deppei*), and by the
strikingly different ventral coloration in adult males (*deppei* is entirely
black ventrally). Superficially *immutabilis* resembles *C. lineatissimus lividus*
which occurs on the coast of Michoacán; this form differs in having a broad

orange-yellow middorsal light area in adults, blue flanks and feet, nine or ten stripes in the young, normally fewer granules around the body (126 to 164 [148.0 $\pm$ 1.24]), and a maximum known snout-vent length of 106 mm.

ECOLOGICAL NOTES.—Individuals of *Cnemidophorus guttatus immutabilis* have been noted only in or close to shaded areas. They inhabit gallery forest along streams, semi-deciduous broad-leaf tropical forest, and dense arid scrub forest. Occasionally they have been observed in association with *Ameiva undulata*. Where *immutabilis* occurs with *Cnemidophorus deppei*, the latter species usually is found only in the more open and sunny situations.

DISTRIBUTION.—This race of *Cnemidophorus guttatus* ranges in coastal lowlands and foothills below elevations of 1000 meters from at least Zihuatanejo, Guerrero, southward across the Plains of Tehuantepec. North of the plains it is known from the Plains of Chivela near the continental divide in the Isthmus of Tehuantepec. At the eastern edge of the Plains of Tehuantepec it apparently intergrades with *C. guttatus flavilineatus*. Northward it probably extends to the Río Balsas, to the north of which it has not been taken and appears to be replaced by *Cnemidophorus lineatissimus lividus*. Locality records (440 specimens) follow.

México: *Guerrero*: Acahuitzotla, UMMZ 119128; Acapulco, UIMNH 3669–73, 38091; 6 km. N of Acapulco, UIMNH 36079–82; UMMZ 104450 (3); 13 km. N of Acapulco, UIMNH 36681–91; Agua del Obispo, UIMNH 36076–7, 36084–5, 36109–10, 36679; 3 km. W of Bajos del Ejido, UMMZ 119129 (13); Buenavista, UIMNH 36083; 5 km. W of Cacalutla, UMMZ 119133 (14); near Chilpancingo, UIMNH 36075; Coyuca, UIMNH 36289, 36698–36707, UMMZ 85416 (4), 105615 (2); 8 km. E of Coyuca, UIMNH 36275; 13 km. N of Coyuca, UIMNH 36692–7; 14 km. NW of Coyuca, UIMNH 36325–35; 2 km. N of El Treinta, UMMZ 119130 (13), 119131 (11), 119132 (11); Garrapatas, UIMNH 36680; Laguna Coyuca, UMMZ 80947 (2); opposite White Friars, UMMZ 84249 (2); Organos, UIMNH 36103–8; Tierra Colorado, UIMNH 36524–58; Xaltinanguis, UIMNH 36078, 36086; Zihuatanejo, UMMZ 84245, 84248 (3). *Oaxaca*: Bahía Escondido, UMMZ 84250 (7); Benito Juarez, UIMNH 8534; 10 km. NW of Camarón, UMMZ 119518; Cerro Quiengola, UIMNH 36569–72; Cerro Tres Cruces, 32 km. SW of Tehuantepec, UIMNH 36559–67, UMMZ 81890; Chacalapa, UKMNH 38241–54; Chivela, UMMZ 67693 (2); 5 km. S of Chivela, UMMZ 114764; Escurano, UIMNH 36573, 36575; between Huilotepec and Tehuantepec, UMMZ 81882 (3), 81883–8, 81892–3; Ixtepec, UIMNH 36674–5; Lagartero, UIMNH 8535; La Venta, UIMNH 39304–15; Mazahua, UMMZ 114763; 5 km. S of Nejapa, UKMNH 44044, 44082; Palmar, UIMNH 36574; 13 km. S of Pachutla, UIMNH 8531–3; Puerto Angel, UIMNH 8536–50; between Quiengola and Tehuantepec, UMMZ 81881 (6), 81889 (3); Río Tequisistlán, UIMNH 39316–24; Salina Cruz, UIMNH 36576–8, UMMZ 81894 (3), 119134 (5); San José Chiltepec, UIMNH 37354; San José Lachiquiri, UIMNH 36568; between Santa Rosa and Tehuantepec, UMMZ 81891 (3); Tehuantepec, UIMNH 36579–36637, 36647–68, 40923, UMMZ 81874 (2), 81875–7, 81878 (5), 81879 (2), 81880 (6); 5 km. WNW of Tehuantepec, UKMNH 33718, 37854–8, UMMZ 112950; 8 km. WNW of Tehuantepec, UKMNH 44083, 44085–8; 13 km. WNW of

Tehuantepec, UKMNH 33714; 16 km. WNW of Tehuantepec, UKMNH 33716; 27 km. W of Tehuantepec, UKMNH 33713; 32 km. W of Tehuantepec, UMMZ 112664; Totolapam, UIMNH 36111-9.

Cnemidophorus guttatus flavilineatus, new subspecies[5]
(Pl. I)

Cnemidophorus guttatus, Burt, 1931, Bull. U. S. Natl. Mus., 154: 66-74 (part).

Cnemidophorus guttatus immutabilis, Smith and Taylor, 1950, Bull. U. S. Natl. Mus., 199: 180 (part).

HOLOTYPE.—University of Michigan Museum of Zoology No. 119465, from Finca Orizaba, about 20 kilometers southwest of Las Cruces, Chiapas, México ($\pm$ 650 meters), collected by Fred G. Thompson, December 22, 1955. Original number, AB 8708.

PARATOPOTYPES.—UMMZ 113825, 119461-119464.

DIAGNOSIS.—A small race of *Cnemidophorus guttatus* with a low number of dorsal granules at midbody (average about 159) and persistent stripes contrasting with a dark dorsal ground color in adults.

DESCRIPTION OF HOLOTYPE.—A large subadult male with a snout-vent length of 110 mm., a tail length (complete) of 228 mm., and a tail/body ratio of 2.08. The scutellation is typical of *guttatus*—three supraoculars, enlarged mesoptychials, and granular postantebrachials. The supraorbital semicircle series extend anteriorly to the posterior edge of the frontal. At midbody there are 142 dorsal granules; there are seven preanal scales and 39 femoral pores.

The top of the head is olive-brown; this color extends onto the rostral, nasals, and postnasals. The upper half of the loreal is black; the preoculars, suboculars, labials, and the lower part of the loreal are bluish cream. The upper surfaces of the limbs and proximal fourth of the tail are unicolor light grayish brown. Posteriorly the tail is bluish gray; the upper surfaces of the hind feet are a darker bluish gray. The chin and ventral surfaces of the forelimbs are light buff; the belly and anterior surfaces of the hind limbs are light blue. The ventral surfaces of the hind limbs and the preanal region are cream; the ventral surface of the tail is cream blending to bluish gray laterally. The dorsal body pattern consists of six separate stripes (laterals, dorsolaterals, and paravertebrals) and an incompletely separated pair of vertebral stripes. At midbody, 32 granules separate the enlarged ventrals from the lateral stripe, which is six granules in width. The flank below the lateral stripe is pale tannish gray and only slightly differentiated

---

[5] From the Latin, *flavus,* yellow, and *lineatus,* lined; used in reference to the broad yellow middorsal stripe in adults.

from the cream lateral stripe. Above the lateral stripe is a deep brownish black lateral dark field ten granules in width. Above this is a cream dorsolateral stripe five granules in width, originating on the posterior superciliary and passing over the insertion of the hind limb to disappear on the proximal dorsolateral surface of the tail. Median to the dorsolateral stripe is a dark field eight granules in width. Anterior to the forelimbs this field is black, but chocolate brown throughout most of the body, fading to light grayish brown posteriorly. Medially is a narrow (three granules in width) paravertebral stripe which is cream anteriorly and brownish cream posteriorly. Between the paravertebral stripes the dark field (26 granules in width at midbody) is deep brownish black. Beginning immediately posterior to the interparietal and extending the length of the body is a broad cream vertebral stripe. On the anterior one-third of the body this stripe is single; then it bifurcates and continues posteriorly as an interconnected pair of vertebral stripes which fuse again above the insertion of the hind limbs. The interspace between the stripes is deep brownish black.

The color in life (of the type series from field notes by Fred G. Thompson): dorsal and lateral light stripes yellow with a slight tinge of green; ground color between dorsal stripes rich chocolate brown; lateral field black; tail slightly greenish; throat orange; belly blue.

DESCRIPTION OF THE SUBSPECIES.—The following description is based on 41 specimens from scattered localities in southwestern Chiapas. The largest male has a snout-vent length of 113 mm.; the largest female, 93 mm. Seven males with snout-vent lengths of more than 100 mm. have tail lengths of 228 to 262 mm. and tail/body ratios of 2.08 to 2.45 (2.27). The smallest juvenile has a snout-vent length of 38 mm.

Scutellation: The dorsal granules vary in number at midbody from 142 to 184 (158.8 ± 1.53). One specimen with 184 and one with 179 are the only ones with more than 173 granules; 55 per cent of the specimens have fewer than 158 granules. The enlarged preanal scales are preceded in the apex of the femora by a varying number of small scales; the number of preanal scales varies from 5 to 10 (7.5). The femoral pores vary in number from 32 to 45 (38.2 ± 0.42). In 12 specimens the supraorbital semicircle series are complete. They extend to the anterior edge of the second supraocular and are continuous with a row of granules separating the first and second supraoculars which, in turn, are continuous with a row of granules between the superciliaries and the supraoculars. In the other specimens the series extend anteriorly only to the posterior edge of the frontal.

Coloration: Juveniles have a black body fading to gray below the lateral stripe, an olive-gray head, and a pale blue tail. The stripes are bold creamy white. There are large cream spots on the forelimbs and narrow cream stripes on the hindlimbs. The vertebral stripe may be single for its entire length, or it may bifurcate somewhere on the anterior one-third of the body. In eight adults and subadults the vertebral stripe is single; in 12 it is bifurcated but interconnected throughout its length; in three the vertebral stripes are entirely separate, and in two of these there is a median row of spots between the vertebral stripes posteriorly; in the others the vertebral stripe is bifurcated and not interconnected. In some subadults there is a faint light stripe on the flank below the lateral stripe. The flanks in adults are pale tannish gray. All individuals have a dark lateral field. Usually the lateral field between the dorsolateral and paravertebral stripes is lighter than the other dark fields. Two individuals have spots developed in the light stripes, but the stripes are continuous and not fragmented into rows of spots. Adult males have a buff colored chin, a black gular collar, and a pale blue belly. In large specimens the blue is darker laterally with scattered white spots.

In life, specimens from the vicinity of Arriaga had a bluish gray lateral stripe, pale cream dorsolateral and paravertebral stripes, and a bright creamy yellow vertebral stripe. The lateral dark field was a rich brownish black; the other dark fields were chocolate brown faintly tinged with pale green. The throat of an adult male was pale orange-buff. A juvenile had pale cream stripes and a bright light blue tail.

ONTOGENETIC CHANGE IN COLOR PATTERN.—The metamorphosis of color pattern in this form does not involve the replacement of stripes with spots (Fig. 16). Apparently the nature of the vertebral stripe (single, bifurcate, or paired) does not change with age. The major change in color pattern is in the pigmentation of the dark fields. The flanks become lighter in old individuals, whereas the lateral dark field remains dark, almost black as in the juveniles. As often as not the dark fields between the paravertebral stripes remain dark, especially anteriorly, while the dorsolateral fields become lighter. In adults the spots and stripes on the limbs vanish. Apparently prior to the disappearance of the light markings on the hind limbs the stripes break into spots.

SEXUAL DIMORPHISM.—Aside from the larger size attained by the males, the only difference between the sexes is in the coloration of the venter. Adult males have a buff throat, black gular collar, and a blue belly; females have an immaculate creamish white venter.

GEOGRAPHIC VARIATION.—The range of this form is divided by the relatively low crest of the Sierra Madre, leaving populations on the Pacific Coast and foothills of the Sierra Madre and in the upper reaches of the Río Cintalapa drainage to the north of the continental divide. There are distinct differences in scutellation between the sample from the coastal lowlands and that from the Cintalapa Valley. In comparison with individuals from the inland localities at elevations of more than 500 meters, specimens from the Pacific lowlands (Arriaga and Tonolá) have more dorsal granules at midbody (165.4 ± 1.72, as compared with 151.3 ± 0.98). Specimens from the lowlands have 33 to 45 (39.0 ± 0.61) femoral pores as compared with 32 to 42 (37.2 ± 0.55) in specimens from Cintalapa. Also, specimens from the lowlands have more preanal scales, 6 to 10 (8.0) as compared with 5 to 9 (7.0) in specimens from Cintalapa. Perhaps there is sufficient genetic discontinuity between the Pacific coastal population and that in the upper Cintalapa Valley to effect such noticeable differences between the populations.

Specimens of *guttatus immutabilis* from Tehuantepec have 179.4 ± 2.29 dorsal granules, 45.2 ± 0.60 femoral pores, and an average of 8.7 preanal scales. In each of these characteristics the Pacific Coast sample of *flavilineatus* is intermediate between typical *immutabilis* and the sample of *flavilineatus* from the Cintalapa Valley. However, all of the Chiapan specimens are alike in color pattern and distinctly different from the majority of *immutabilis* from Tehuantepec (see discussion of geographic variation in *C. guttatus immutabilis*).

Three specimens from Tapanatepec and one from San Juanico in extreme southeastern Oaxaca have 146, 147, 168, and 168 dorsal granules at midbody, and 38, 42, 45, and 46 femoral pores. In two the vertebral stripe is single, and in two it is bifurcate. Although two are large males (snout-vent lengths of 112 and 113 mm.), neither shows any fragmentation of the stripes into spots, as is characteristic of *C. g. immutabilis* a few kilometers to the west on the Plains of Tehuantepec. Consequently, these specimens are assigned to *flavilineatus*.

COMPARISON WITH OTHER FORMS.—This form differs from the other races of *guttatus* in having fewer dorsal granules at midbody (158.8 ± 1.53, as compared with 176.8 ± 1.54 in *immutabilis* and 199.3 ± 1.11 in *guttatus*) and in having a dorsal color pattern of persistent light stripes on a dark brown dorsum, as compared with rows of spots or stripes partly fragmented into spots in the other races.

From *C. d. deppei*, which occurs in the same area, *flavilineatus* may be distinguished by its greater maximum size (113 mm. as compared with 80 mm.), greater number of dorsal granules (more than 140 as compared with fewer than 120), and in having a broad vertebral light stripe. Superficially *flavilineatus* resembles *C. lineatissimus lividus* from Michoacán. There are only minor differences in scutellation; *flavilineatus* has slightly more dorsal granules at midbody than *lividus* (158.8 ± 1.53 as compared with 148.0 ± 1.24). Both forms have a persistent broad vertebral light stripe, but *flavilineatus* has only one pair of dorsolateral stripes, whereas *lividus* has two.

ECOLOGICAL NOTES.—Individuals of this race have been collected in dense scrub forest and semi-deciduous broad-leaf tropical forest environments on the coastal lowlands and foothills, and in somewhat drier scrub forest in the Cintalapa Valley.

DISTRIBUTION.—*Cnemidophorus guttatus flavilineatus* is known only from the upper Cintalapa Valley and from the coastal lowlands and foothills of southwestern Chiapas and extreme southeastern Oaxaca. Although the Río Cintalapa is in the Gulf drainage and is a tributary of the Río Grijalva (via the Río de la Venta), specimens of *flavilineatus* are unknown from the broad Grijalva Valley in central and eastern Chiapas. Apparently in the more humid coastal lowlands to the east of Pijijiapan it is replaced by *Ameiva undulata*. Two specimens (crest above Arriaga and Buena Vista) are from elevations of about 1000 meters in the Sierra Madre. Possibly the population is continuous from the Pacific lowlands across the Sierra Madre to the Cintalapa Valley; however, the differences in the present samples from these localities suggest a minimum of continuity (see discussion of geographical variation in *flavilineatus*). Apparently *flavilineatus* intergrades with *immutabilis* at the eastern edge of the Plains of Tehuantepec. Locality records (94 specimens) follow.

México: *Chiapas*: Arriaga, UMMZ 88379, 94828–37, 94889; 12 km. N of Arriaga, UMMZ 114760 (4); 16 km. N of Arriaga, UMMZ 114759; Buena Vista, UMMZ 113827; Cintalapa, UMMZ 99833 (9), 99834; 16 km. NE of Cintalapa, UMMZ 118726; crest above Arriaga, UMMZ 94875; Finca Orizaba, UMMZ 113825, 119461–5; Hacienda Monserrate, UMMZ 102222 (3); Pijijiapan, UMMZ 119513–5; Rancho San Bartolo, UIMNH 8551–61, 8571–83; San Ricardo, UIMNH 36128; Tonolá, UIMNH 36120–3, 36638–46, UMMZ 88378 (2), 119516; 10 km. NW of Tonolá, UKMNH 43914, 43922, 43926–7, 44084, 44089–90. *Oaxaca*: San Juanico, UMMZ 113826; Tapanatepec, UMMZ 84493 (2), 118725.

DISCUSSION

From the preceding descriptions of variation, ontogenetic change in color pattern, distribution, and ecology, certain conclusions concerning the phylogenetic relationships of the species in this group may be drawn. The three species comprising the *deppei* group of the genus *Cnemidophorus* belong to a single phyletic line, which, according to Burt (1931: 255), arose from a stock derived from the South American *lemniscatus* group. Not all workers hold with Burt's interpretation of the relationships and evolutionary history of the genus. Until detailed studies have been completed on all of the species groups in the genus no substantial analysis of the interrelationships of the various groups can be made. Consequently, we shall limit our remarks to the *deppei* group.

Burt (*loc. cit.*) suggested that the *deppei* group probably did not evolve before the Late Miocene; furthermore, he explained the differentiation of *deppei* and *guttatus* (the two species recognized by him) by the isolation of stocks by a sea portal at the Isthmus of Tehuantepec in southern México. The recognition of three species and ten races in this group, instead of two species and three races as recognized by Burt, necessitates a reinterpretation of the phylogenetic history of the group.

Before discussing the phylogenetic history of the group, it is necessary to ascertain the relationships of the species. *Cnemidophorus deppei* differs from the others in having black venters in adult males, generally persistent stripes, and fewer dorsal granules. On the other hand, *guttatus* and *lineatissimus* appear to be more closely related in that they both have blue bellies and black gular collars in adult males, fragmentation of stripes into spots or vertical bars, and a greater number of dorsal granules than *deppei*.

Since the juveniles of all forms except *guttatus guttatus* are striped, it may be assumed that the ancestral *deppei* stock had a pattern of stripes, and that those forms in which the stripes are either completely or partially replaced by spots or vertical bars have evolved from a striped form. Juveniles of *C. guttatus guttatus* have only a lateral stripe; adults have rows of spots which develop without the fragmentation of stripes. In this respect, and possibly with respect to its greater number of dorsal granules, *guttatus guttatus* appears to be the most advanced form in the group.

There appears to be a close correlation between color pattern and habitat in these lizards. The adults of *deppei* are striped; these lizards live in open, often grassy, habitats. The spotted forms or those with vertical bars or other fragmentation of the lined pattern (*guttatus* and *lineatissimus*) inhabit shaded areas. Apparently there is selection for the spotted pattern in shaded habitats and for the striped pattern in open environments. We do not have data to determine if the striped juveniles of *guttatus* and *lineatissimus* have habitat preferences different from their spotted adults.

Duellman (1958) presented a synopsis of Middle American paleogeography in which he summarized geological, paleontological, and biogeographical evidence suggesting that if a seaway existed at the Isthmus of Tehuantepec, it must have been present only during the Lower Pliocene, and that even if a seaway did not exist, climatic fluctuations during the Cenozoic probably caused considerable ecological change in the isthmian region, so as to provide both barriers to and highways for animal dispersal. In attempting to reconstruct the phylogenetic history of the group we may assume that prior to the existence of a seaway or other comparable barrier at the Isthmus of Tehuantepec, a *deppei* group prototype ranged from at least northern Central America northward onto the Gulf and Pacific lowlands of México. The presence of a subsequent barrier at the Isthmus of Tehuantepec would have served to isolate three populations of the *deppei* prototype—a *deppei* stock in northern Central America, a *guttatus* stock on the Gulf lowlands of what is now Veracruz, and a *lineatissimus* stock on the Pacific lowlands of México. Subsequent to a period of isolation and differentiation, and following the disappearance of a barrier at the Isthmus of Tehuantepec, the *guttatus* stock, which probably most closely resembled the present race *flavilineatus,* dispersed southward across the Isthmus of Tehuantepec to invade the Pacific lowlands of Chiapas, Oaxaca, and Guerrero. This species lives in shaded environments, but not in rainforest. Climatic fluctuation during the Late Pliocene and Pleistocene may have produced effects resulting in the restriction of the lizard's environment, and thus serving to isolate populations of *guttatus* on the Atlantic lowlands (*guttatus*), Pacific lowlands west of the Isthmus of Tehuantepec (*immutabilis*), and Pacific lowlands (possibly only in the Cintalapa Valley) east of the Isthmus of Tehuantepec (*flavilineatus*).

The *lineatissimus* stock developed on the Pacific lowlands of México, where it lived in shaded forest environments similar to those inhabited by *guttatus*. With the dispersal of *guttatus* onto the Pacific lowlands, *lineatissimus* may have been exterminated south of the Río Balsas. Perhaps *lineatissimus* had moved to the region north of the Río Balsas before the immigration of *guttatus* into Oaxaca and Guerrero. A population of *lineatissimus* became isolated in the Tepalcatepec Valley and subsequently differentiated into *exoristus*. Later this form invaded the lowlands of Colima via the Ahuijullo Depression to differentiate there into *lineatissimus*. To the north along the coastal regions of Jalisco and Nayarit the *lineatissimus* stock differentiated into *duodecemlineatus* and to the south along the coast of Michoacán, into *lividus*.

Cnemidophorus deppei dispersed northward and westward from northern Central America in dry and open environments. Consequently, it was not in competition with either *guttatus* or *lineatissimus*. A population of *deppei* differentiated in the Balsas-Tepalcatepec Valley to form *infernalis;* another population became isolated in open environments in the Yucatán Peninsula and developed into *cozumelus*.

The distribution patterns of these lizards indicate that considerable climatic fluctuation must have occurred since the species differentiated. This is particularly true of *deppei,* for the relict populations of *cozumelus*

on Isla de Cozumel, Isla Mujeres, and especially in the savannas of El
Petén, Guatemala, suggest that savannas or other non-evergreen forest
environments were once much more extensive in that region than they are
now. On the other hand, the presence of a population of *lineatissimus* in
gallery forests in the arid Tepalcatepec Valley indicates that more mesic
environments formerly were more widespread in that area.

SUMMARY

The examination of 2300 specimens of *Cnemidophorus* of the *deppei*
group has revealed the presence of three species and ten races. The species
differ from one another in the relative size of the dorsal granules, the size
of the body, color pattern, and ontogenetic change in color pattern. Two
of the species, *guttatus* and *lineatissimus*, are allopatric and inhabit shaded
areas in arid and subhumid environments; the third species, *deppei*, occurs
sympatrically with the others, but inhabits open situations. The ranges of
the species, particularly *deppei* with its relict populations in the savannas
of Guatemala and on Isla de Cozumel, appear to be discontinuous. The
present distributions probably are the result of climatic shifts during the
Pleistocene.

Cnemidophorus deppei consists of three races—*deppei* from Costa Rica
to Michoacán and Veracruz, *cozumelus* in the Yucatán Peninsula, and
infernalis in the Balsas-Tepalcatepec Valley; *guttatus* also consists of three
races—*guttatus* in Veracruz, *immutabilis* on the Pacific lowlands of Guerrero
and Oaxaca, and *flavilineatus* in the Cintalapa Valley and on the Pacific
lowlands of Chiapas. *Cnemidophorus lineatissimus* is made up of four races
—*lividus* on the coast of Michoacán, *lineatissimus* on the lowlands of
Colima, *duodecemlineatus* on the lowlands of Jalisco and Nayarit, and
exoristus in the Tepalcatepec Valley.

Our investigations of the *deppei* group have revealed several problems
worthy of future study. The most notable systematic problems to be solved
are: (1) The status of the 10-lined *"deppei"* in Veracruz (see discussion of
variation of *Cnemidophorus deppei deppei*), (2) The amount of variation
of color pattern in living specimens of *deppei* from the coast of Chiapas,
where the scutellation is highly variable, (3) The relationships of the
populations of *lineatissimus* in Colima (see discussion of geographic variation in *Cnemidophorus lineatissimus*). Certain questions concerning distribution can be answered only after additional field work; for example,
collections are needed from the savannas of Tabasco and the southern part
of the Yucatán Peninsula to determine the distribution of *deppei* in that
region.

LITERATURE CITED

BURT, CHARLES E.
1931 A study of the teiid lizards of the genus *Cnemidophorus* with special reference to their phylogenetic relationships. U. S. Natl., Mus. Bull., 154: i–viii, 1–286.

COPE, EDWARD D.
1877 Tenth contribution to the herpetology of tropical America. Proc. Amer. Philos. Soc., 17: 85–98.
1892 A synopsis of the species of the teid genus *Cnemidophorus*. Trans. Amer. Philos. Soc., 17: 27–52, Pls. 6–13.
1894 Third addition to a knowledge of the Batrachia and Reptilia of Costa Rica. Proc. Acad. Nat. Sci. Philadelphia, Pp. 194–206.
1900 Crocodilians, lizards, and snakes of North America. Ann. Rept. U. S. Natl. Mus. for 1898, Pp. 151–1294.

DAVIS, WILLIAM B., AND HOBART M. SMITH
1953 Lizards and turtles of the Mexican state of Morelos. Herpetologica, 9: 100–108.

DUELLMAN, WILLIAM E.
1954 The amphibians and reptiles of Jorullo Volcano, Michoacán, Mexico. Occ. Papers Mus. Zool. Univ. Michigan, 560: 1–24, Pl. 1.
1955 A new whiptail lizard, genus *Cnemidophorus*, from Mexico. *Ibid.*, 574: 1–7.
1958 A monographic study of the colubrid snake genus *Leptodeira*. Bull. Amer. Mus. Nat. Hist., 114: 1–151, Pls. 1–31.

GADOW, HANS
1903 Evolution of the colour-pattern and orthogenetic variation in certain Mexican species of lizards, with adaptation to their surroundings. Proc. Royal Soc. London, 72: 109–125.
1905 The distribution of Mexican amphibians and reptiles. Proc. Zool. Soc. London, Pp. 191–244.
1906 A contribution to the study of evolution based upon the Mexican species of *Cnemidophorus*. *Ibid.*: 277–375.

HALLOWELL, EDWARD
1860 Report upon the Reptilia of the North Pacific Exploring Expedition, under the command of Capt. John Rodgers, U.S.N. Proc. Acad. Nat. Sci. Philadelphia, Pp. 480–510.

HARTWEG, NORMAN, AND JAMES OLIVER
1937 A contribution to the herpetology of the Isthmus of Tehuantepec. II. The teiids of the Pacific slope. Occ. Papers Mus. Zool. Univ. Michigan, 359: 1–8.

LEWIS, THOMAS H.
1956 A new lizard of the genus *Cnemidophorus* from Nayarit, Mexico. Chicago Acad. Sci., Nat. Hist. Misc., 156: 1–5.

LOWE, CHARLES H., JR., AND RICHARD G. ZWEIFEL
1952 A new species of whiptailed lizard (genus *Cnemidophorus*) from New Mexico. Bull. Chicago Acad. Sci., 9: 229–47, Pl. 1.

PETERS, JAMES A.
1954 The amphibians and reptiles of the coast and coastal sierra of Michoacán, Mexico. Occ. Papers Mus. Zool. Univ. Michigan, 554: 1–37.

SCHMIDT, KARL P., AND FREDERICK A. SHANNON
    1947 Notes on amphibians and reptiles of Michoacán, Mexico. Fieldiana: Zool.,
        31 (9): 63–85.
SMITH, HOBART M.
    1939 Notes on Mexican reptiles and amphibians. Field Mus. Nat. Hist., Zool. Ser.,
        24 (4): 15–35.
SMITH, HOBART M., AND EDWARD H. TAYLOR
    1950 An annotated checklist and key to the reptiles of Mexico exclusive of the
        snakes. U. S. Natl. Mus. Bull., 199: i–v, 1–253.
WIEGMANN, A. F. A.
    1834 Herpetologia Mexicana. Berlin, Pp. 1–54, Pls. 1–10.
ZWEIFEL, RICHARD G.
    1959 Variation in and distribution of lizards of western Mexico related to *Cnemido-
        phorus sacki*. Bull. Amer. Mus. Nat. Hist., 117 (2): 57–116, Pls. 43–49.

Accepted for publication August 7, 1959

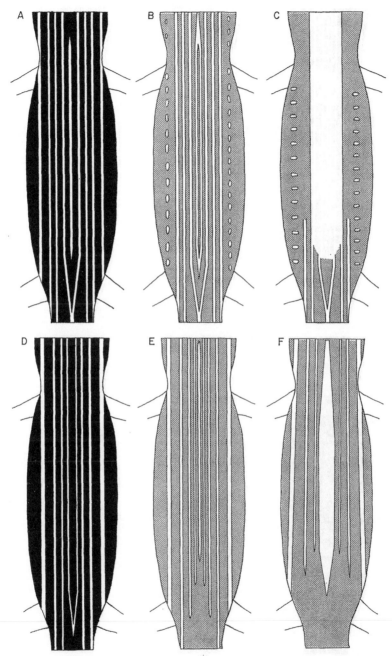

FIG. 11. Ontogenetic change in color pattern in *Cnemidophorus deppei deppei* (A–C) and *Cnemidophorus deppei infernalis* (D–F).

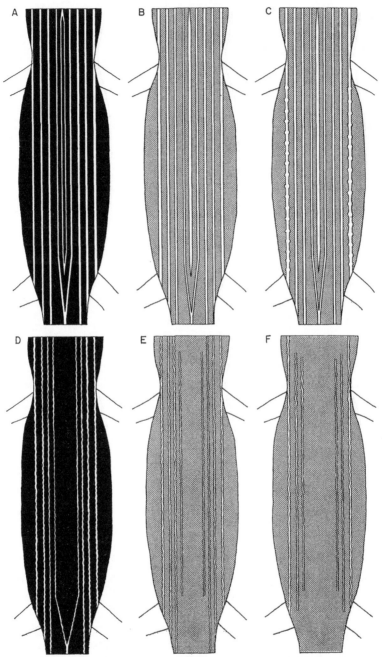

FɪG. 12. Ontogenetic change in color pattern in *Cnemidophorus* cf. *deppei deppei* from Lerdo de Tejada, Veracruz (A–C) and *Cnemidophorus deppei cozumelus* (D–F).

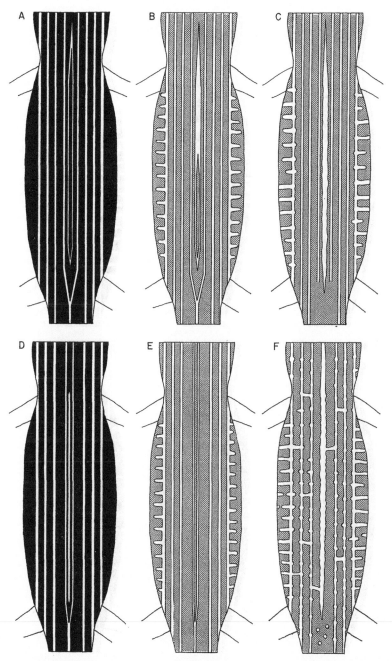

FIG. 13. Ontogenetic change in color pattern in *Cnemidophorus lineatissimus lineatissimus* (A–C) and *Cnemidophorus lineatissimus exoristus* (D–F).

DUELLMAN AND WELLMAN

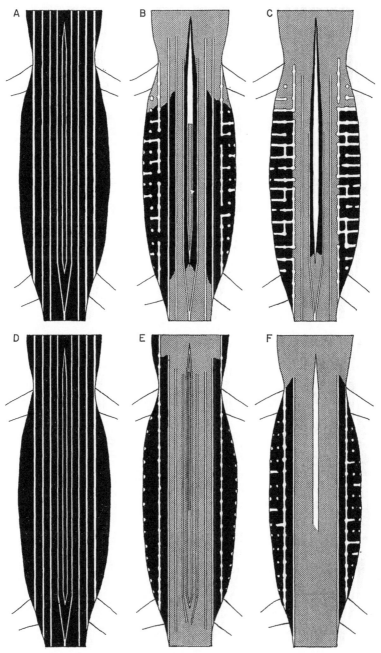

Fig. 14. Ontogenetic change in color pattern in *Cnemidophorus lineatissimus lividus* (A–C) and *Cnemidophorus lineatissimus duodecemlineatus* (D–F).

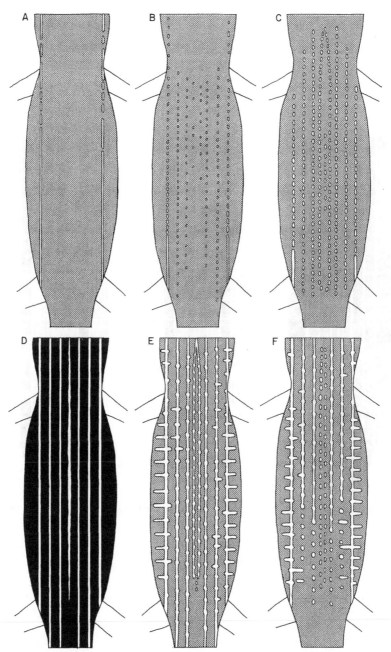

FIG. 15. Ontogenetic change in color pattern in *Cnemidophorus guttatus guttatus* (A–C) and *Cnemidophorus guttatus immutabilis* (D–F).

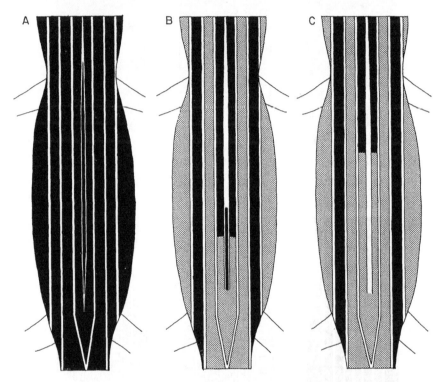

Fig. 16. Ontogenetic change in color pattern in *Cnemidophorus guttatus flavilineatus* (A–C).

PLATE I

TOP: Holotype of *Cnemidophorus deppei infernalis* (UMMZ 114783).
NEXT TO TOP: Holotype of *Cnemidophorus lineatissimus exoristus* (UMMZ 119338).
NEXT TO BOTTOM: Holotype of *Cnemidophorus lineatissimus lividus* (UMMZ 119472).
BOTTOM: Holotype of *Cnemidophorus guttatus flavilineatus* (UMMZ 119465).

Printed and bound by CPI Group (UK) Ltd, Croydon, CR0 4YY

13/04/2025

14656510-0001